Thermal Energy Systems

The text provides in-depth knowledge about recent advances in solar collector systems, photovoltaic systems, the role of thermal energy systems in buildings, phase change materials, geothermal energy, biofuels, and thermal management systems for EVs in social and industrial applications. It further aims toward the inclusion of innovation and implementation of strategies for CO_2 emission reduction through the reduction of energy consumption using conventional sources.

This book:

- Presents the latest advances in the field of thermal energy storage, solar energy development, geothermal energy, and hybrid energy applications for green development.
- Highlights the importance of innovation and implementation of strategies for CO_2 emission reduction through the reduction of energy consumption using sustainable technologies and methods.
- Discusses design development, life cycle assessment, modelling and simulation of thermal energy systems in detail.
- Synergize exploration related to the various properties and functionalities through extensive theoretical and numerical modelling present in the energy sector.
- Explores opportunities, challenges, future perspectives and approaches toward gaining sustainability through renewable energy resources.

The text discusses the fundamentals of thermal energy and its applications in a comprehensive manner. It further covers advancements in solar thermal and photovoltaic systems. The text highlights the contribution of geothermal energy conversion systems to sustainable development. It showcases the design and optimization of ground source heat pumps for space conditioning and presents modelling and simulation of the thermal energy systems for design optimization. It will serve as an ideal reference text for senior undergraduate, graduate students and academic researchers in the fields of mechanical engineering, environmental engineering and energy engineering.

Advances in Manufacturing, Design and Computational Intelligence Techniques

Series Editor: Ashwani Kumar

The book series editor is inviting edited, reference and text book proposal submission in the book series. The main objective of this book series is to provide researchers a platform to present state of the art innovations, research related to advanced materials applications, cutting edge manufacturing techniques, innovative design and computational intelligence methods used for solving nonlinear problems of engineering. The series includes a comprehensive range of topics and its application in engineering areas such as additive manufacturing, nanomanufacturing, micromachining, biodegradable composites, material synthesis and processing, energy materials, polymers and soft matter, nonlinear dynamics, dynamics of complex systems, MEMS, green and sustainable technologies, vibration control, AI in power station, analog-digital hybrid modulation, advancement in inverter technology, adaptive piezoelectric energy harvesting circuit, contactless energy transfer system, energy efficient motors, bioinformatics, computer aided inspection planning, hybrid electrical vehicle, autonomous vehicle, object identification, machine intelligence, deep learning, control-robotics-automation, knowledge based simulation, biomedical imaging, image processing and visualization. This book series compiled all aspects of manufacturing, design and computational intelligence techniques from fundamental principles to current advanced concepts.

Thermal Energy Systems: Design, Computational Techniques, and Applications (Ashwani Kumar)
Edited by Ashwani Kumar, Varun Pratap Singh, Chandan Swaroop Meena, Nitesh Dutt

Thermal Energy Systems
Design, Computational Techniques, and Applications

Edited by
Ashwani Kumar
Varun Pratap Singh
Chandan Swaroop Meena
Nitesh Dutt

CRC Press is an imprint of the
Taylor & Francis Group, an **informa** business

Front cover image: metamorworks/Shutterstock

First edition published 2023
by CRC Press
6000 Broken Sound Parkway NW, Suite 300, Boca Raton, FL 33487-2742

and by CRC Press
4 Park Square, Milton Park, Abingdon, Oxon, OX14 4RN

CRC Press is an imprint of Taylor & Francis Group, LLC

© 2023 selection and editorial matter, Ashwani Kumar, Varun Pratap Singh, Chandan Swaroop Meena and Nitesh Dutt; individual chapters, the contributors

Reasonable efforts have been made to publish reliable data and information, but the author and publisher cannot assume responsibility for the validity of all materials or the consequences of their use. The authors and publishers have attempted to trace the copyright holders of all material reproduced in this publication and apologize to copyright holders if permission to publish in this form has not been obtained. If any copyright material has not been acknowledged, please write and let us know so we may rectify in any future reprint.

Except as permitted under U.S. Copyright Law, no part of this book may be reprinted, reproduced, transmitted, or utilized in any form by any electronic, mechanical, or other means, now known or hereafter invented, including photocopying, microfilming, and recording, or in any information storage or retrieval system, without written permission from the publishers.

For permission to photocopy or use material electronically from this work, access www.copyright.com or contact the Copyright Clearance Center, Inc. (CCC), 222 Rosewood Drive, Danvers, MA 01923, 978-750-8400. For works that are not available on CCC please contact mpkbookspermissions@tandf.co.uk

Trademark notice: Product or corporate names may be trademarks or registered trademarks and are used only for identification and explanation without intent to infringe.

Library of Congress Cataloguing-in-Publication Data
Names: Kumar, Ashwani, 1989-editor.
Title: Thermal energy systems : design, computational techniques, and applications / edited by Ashwani Kumar, Varun Pratap Singh, Chandan Swaroop Meena, Nitesh Dutt.
Description: First edition. | Boca Raton : CRC Press, [2023] |
Series: Advances in manufacturing, design and computational intelligence techniques | Includes bibliographical references and index.
Identifiers: LCCN 2022059386 (print) | LCCN 2022059387 (ebook) |
ISBN 9781032392936 (hbk) | ISBN 9781032498508 (pbk) | ISBN 9781003395768 (ebk)
Subjects: LCSH: Heat engineering.
Classification: LCC TJ255 .T45 2023 (print) | LCC TJ255 (ebook) | DDC 621.402--dc23/eng/20230111
LC record available at https://lccn.loc.gov/2022059386
LC ebook record available at https://lccn.loc.gov/2022059387

ISBN: 978-1-032-39293-6 (hbk)
ISBN: 978-1-032-49850-8 (pbk)
ISBN: 978-1-003-39576-8 (ebk)

DOI: 10.1201/9781003395768

Typeset in Sabon
by MPS Limited, Dehradun

Contents

Aim and Scope	ix
Preface	xi
Acknowledgement	xv
About the Editors	xvii
List of contributors	xix

1 Introduction to Thermal Energy Resources and Their Smart Applications   1

ARIJIT KUNDU, ASHWANI KUMAR, NITESH DUTT, CHANDAN SWAROOP MEENA, AND VARUN PRATAP SINGH

2 Thermal Energy Storage: Opportunities, Challenges and Future Scope   17

ASHOK K. DEWANGAN, SYED QUADIR MOINUDDIN, MURALIMOHAN CHEEPU, SANJEEV K. SAJJAN, AND ASHWANI KUMAR

3 Introduction and Fundamentals of Solar Energy Collectors   29

MADHUR CHAUHAN, BIPASA PATRA, AND PRANAV CHARKHA

4 Optimization of Solar Collector System Based on Different Nanofluids   41

OSAMA KHAN, S. MOJAHID UL ISLAM, TAUSEEF HASSAN, HOZAIFA AHMAD, MD ADIB UR RAHMAN, AND ASHOK K. DEWANGAN

5 Advancements in Solar Thermal and Photovoltaic System   59

BIPASA PATRA AND PRAGYA NEMA

vi Contents

6 Thermal Energy Applications in Net-Zero Energy Buildings   73

MRITYUNJAI VERMA AND ASHISH KARN

7 Modelling and Simulation of Thermal Energy System for Design Optimization   103

ARIJIT KUNDU, ASHWANI KUMAR, NITESH DUTT, VARUN PRATAP SINGH, AND CHANDAN SWAROOP MEENA

8 Thermal Efficiency Enhancement of Solar Still Using Fins with PCM   141

NAVEEN SHARMA, NOUSHAD SHAIK, VIVEK KUMAR, AND MUKESH KUMAR

9 Thermal and Electrical Management of a Solar PV/T System with and without PCM   155

ANKIT DEV, RAVI KUMAR, R.P. SAINI, AND ADITYA KUMAR

10 Second Law Analysis of Desiccant Cooling-Based Thermal Systems   169

D.B. JANI

11 Analysis of Optimum Operating Parameters for Ground Source Heat Pump System for Different Cases of Building Heating and Cooling Mode Operations   183

T. SIVASAKTHIVEL, VIKAS VERMA, RAHUL TARODIYA, CHANDAN SWAROOP MEENA, VARUN PRATAP SINGH, AND RAJESH KUMAR

12 Lithium-Ion Battery Thermal Management Systems with Different Mediums and Techniques for Electric-Driven Vehicles   219

SAURAV SIKARWAR, RAJESH KUMAR, ASHOK YADAV, NITESH DUTT, VIKAS VERMA, AND T. SIVASAKTHIVEL

13 Performance Evaluation of Diesel Engine with Fuels Prepared from Hydrogen and Nanoparticle Blended Biodiesel by Varying Injection Pressure   231

MOHAMMAD ASHAD GHANI NASIM, MOHD PARVEZ, OSAMA KHAN, GULAM HASNAIN WARSI, MD HASSAAN, AND ASHOK K. DEWANGAN

14 Effect of Antioxidant *Psidium guajava* Extract on the Stability of Oxidation of Various Biodiesels 249

MEETU SINGH, NEERJA, AMIT SARIN, DEEPTAM TRIVEDI,
SUJEET KESHARVANI, ANJALI AGRAWAL, AND GAURAV DWIVEDI

Index 275

Aim and Scope

The global concern of the 21st century is climate change and global warming. The main cause of these concerns is the increasing greenhouse gas emissions into the environment. Since the beginning of the industrial era, greenhouse gas emissions are increasing year by year. Energy consumption is proportional to the increasing population growth and inversely proportional to the global environment's health. These consumption rates have bought an increase in global warming potential, GHG emissions, where CO_2 is the major contributor. *Energy crisis is another add-on factor towards these increasing dilemmas of environmental effects and consequent increase in fuel price which do have major possible solution sources and efficient use of technologies within the system to reduce the impact of increased demand.* Researchers and policymakers have had a keen interest towards changing policies in these responses where inter-governmental panel on climate change (IPCC) used the global energy potential (GEP) index to standardize variable gas input, CDM established in assessment to Kyoto Protocol to make the developing nations to attain sustainable development and developed nations to have convergence with the qualified GHG emission. The "20-20-20" law is enforced by European Union as climate and energy package where first 20% is for consumption reduction, next 20% is for CO_2 emission reduction and with last 20% is assigned to energy from renewables. Net Zero Energy Buildings (NZEB) is highlighted in the book as a major renewable energy application.

India has committed to increase its share of renewables by 450 GW by 2030. The potential of solar energy for India is inevitable. Energy crises have bought in the mindset of developing nations to procure a sustainable and reliable domestic source of energy. In this scenario, a book titled *Thermal Energy Systems: Design, Computational Techniques, and Applications* is the time of requirement to compile all details of the latest research finding in one place. In this book, major attention will be on the basic applications of *solar energy, geothermal energy and hybrid energy.* Researchers are continuously working on hybrid energy applications, with the evolution of "almost zero energy" concept giving the uprise to nearly zero energy buildings (nZEB). *Solar photovoltaics are the fastest-growing electricity source.*

Though the foundation of the industrial era is set upon fossil resources, a rigorous and conducive change towards renewables is much anticipated to negotiate the crisis sooner. Energy-efficient products such as LEDs, CFLs could be substituted to avert the energy crisis, if this is added with a sensor-based lighting system it could work wonders. *Easy renewable in situ grid options must be available by using solar and thermal energies.*

A clear and responsible input from both developed and developing countries towards lowering the GHG emissions, like the one taken under the Paris agreement is consistently persuaded. Renewable energy is powering a clean energy revolution. *Basic renewable energy generators solar energy, geothermal energy and applications of hybrid energies application is the central theme of the book.* They are the main contributor to sustainable development.

Editors
Dr. Ashwani Kumar
Dr. Varun Pratap Singh
Dr. Chandan Swaroop Meena
Dr. Nitesh Dutt

Preface

The book *Thermal Energy Systems: Design, Computational Techniques and Applications* provides in-depth knowledge about recent research trends in solar energy development, thermal energy, geothermal energy and hybrid energy for social and industrial applications. The book aims towards the inclusion of innovation and implementation of strategies for CO_2 emission reduction through the reduction of energy consumption using conventional sources (fossil fuels, coals and hydropower). *The main focus of the book has been given to the application of solar and thermal energy with the application of hybrid energy, a one-step solution to mitigate carbon footprint and greenhouse gas emissions and moving the energy consumption towards a sustainable future.*

The most commonly used renewable energy is solar energy, due to its cost-effectiveness and ease of availability. It can be utilized through roof-type solar photovoltaic (PV) panels or as a Building Integrated Photovoltaic (BIPV) system. *Heat pumps are used in harnessing geothermal energy resources to generate electricity.* Large solar grids, most often used by utilities to provide power to a grid, range from 100 kilowatts to several megawatts. Geothermal and solar, hybrid energy is powering a clean energy revolution. Geothermal energy can heat, cool and generate electricity: Application of Geothermal energy depends on the resource and technology chosen—heating and cooling buildings through geothermal heat pumps, generating electricity through geothermal power plants and heating structures through direct-use. In the proposed book, major attention will be on the basic applications of *solar energy, geothermal energy and hybrid energy*.

Chapter 1 states that Thermal Energy has been utilized as the most fundamental type of energy for fundamental tasks including heating, cooking and boiling water from the beginning of the human species. Later several additional natural and man-made sources, such as electrical heating, solar energy, geothermal energy, nuclear heat from the fusion and fission process, etc., have come into existence over time and have been utilized by humankind. In continuation, Chapter 2 aims to provide insight into challenges in storing thermal energy, technologies (sensible, latent heat, thermo-chemical) required for TES, and materials (phase change materials) required for TES devices and performances. In addition, the chapter also focuses on TES for solar systems that can revolutionize the energy industries and Chapter 3 discusses solar energy collectors in detail.

Chapter 4 deals with exploring multiple nanofluid combinations for cooling the flat plate collector by varying the operating conditions. Analytical hierarchy process (AHP) is used to weigh the operating parameters and TOPSIS method is used to rank the nanofluids circulating in the flat plate collector. Nanofluids after analysis are ranked in descending order based on the acquired closeness coefficient score: Graphene oxide > Zinc oxide > Silicon oxide > Deionized Water. Chapter 5 summarizes solar cell materials including crystalline, thin film technology, solar PV concentrated, polymer and organic configuration, hybrid configuration, dye-sensitized, etc.

Chapter 6 highlights the important theme of the book, i.e., Net-Zero Energy Buildings. A net-zero energy building, or nZEB, is a matrix construction with exceptional embodied energy. nZEB ensures that its fundamental power consumption is balanced by maintaining that the proportion of electricity production delivered through a meter or into any other energy net equals the amount of primary power required to nZEB via power systems. As a reason, a net-zero energy building will only generate energy when the conditions are favourable and will rely on given energy at all other times. If great efficacy in buildings is to be attained by the use of generally accepted and extensively defined indicators, a large decrease in energy-related carbon emissions must be legislated for virtually zero-energy buildings (nZEB).

Chapters 7 and 8 deal with the modelling and simulation of thermal energy systems for thermal efficiency enhancement. Authors have explored the influence of hollow and phase change material (PCM) filled copper fins on the thermal effectiveness of pyramid solar stills (PSS). In continuation, Chapters 9 and 10 deal with lowering the PV surface temperature at various solar irradiations utilizing active cooling using various heat transfer fluids and explains desiccant cooling-based thermal systems. The PV panel integrated with active cooling is known as PV/T system. A metal container filled with Paraffin wax PCM was also used along with active cooling. A total of five fluids, i.e., water and four different nanofluids were used to evaluate the temperature and electrical output of the newly designed PV/T-PCM system.

In Chapter 11, the authors have studied the five different cases of buildings are considered for analysis: (i) building requires only heating, (ii) building requires only cooling, (iii) building heating load is higher than cooling load, (iv) building heating load equal to cooling load and (v) building heating load is less than cooling load. The study focuses on how GHX is able to meet different cases of cooling and heating loads and to select the optimum parameters which will reduce the required heat exchanger length. To achieve this objective, Taguchi method has been employed.

Chapter 12 highlights the emerging lithium-ion battery thermal management systems. The lithium-ion battery (LiB) is a winner among all traction batteries for the application in EVs and (hybrid electric vehicles) HEVs. Furthermore, LiB gives the best efficiency within the limited temperature range from 15°C to 35°C. To maintain the temperature within the optimum range, the battery thermal management system plays a vital role. The battery thermal management systems (BTMSs) generally utilize the air, liquid & phase change material (PCM) as working. The selection of BTMS type is dependent on the rating of a battery pack (BP).

Chapters 13 and 14 deal with biofuels. Biodiesel is gaining popularity as an alternative, biodegradable, non-toxic and renewable energy source. The stability of oxidation is an important factor in determining the self-life of any biodiesel, and several antioxidants are utilized to avoid oxidative degradation. In this study, the antioxidative potential of *Psidium guajava* extract (PGE) was investigated in order to improve the oxidation stability of three biodiesels: *Jatropha* biodiesel (JBD), *Pongamia* biodiesel (PBD) and *Tectona grandis* biodiesel (TGBD). Chapters 1–14 presented in the book make it an ideal book for research scholars, upper-level undergraduate and graduate students, engineers, technologists and energy scientists working in the area of thermal energy storage, design and distribution.

Editors
Dr. Ashwani Kumar
Dr. Varun Pratap Singh
Dr. Chandan Swaroop Meena
Dr. Nitesh Dutt

Acknowledgement

We express our gratitude to CRC Press (Taylor & Francis Group) and the editorial team for their suggestions and support during the completion of this book. We are grateful to all contributors and reviewers for their illuminating views on each book chapter presented in the book *Thermal Energy Systems: Design, Computational Techniques and Applications.*

"This book is dedicated to all engineers, researchers and academicians..."

About the Editors

Ashwani Kumar received a PhD (Mechanical Engineering) in Mechanical Vibration and Design. He is currently working as a Senior Lecturer, Mechanical Engineering (Gazetted Officer Group B) at Technical Education Department, Uttar Pradesh (Under the Government of Uttar Pradesh), Kanpur, India, since December 2013. He worked as an Assistant Professor in the Department of Mechanical Engineering, Graphic Era University, Dehradun, India, from July 2010 to November 2013. He has 12 years of research and academic experience in mechanical and materials engineering. He is the series editor of the book series, *Advances in Manufacturing, Design and Computational Intelligence Techniques* and *Renewable and Sustainable Energy Developments* published by CRC Press, Taylor & Francis, USA. He is the Editor-in-Chief for *International Journal of Materials, Manufacturing and Sustainable Technologies (IJMMST)* and the Associate Editor for*International Journal of Mathematical, Engineering and Management Sciences (IJMEMS)* Indexed in ESCI/Scopus and DOAJ. He is an editorial board member of four international journals and acts as a review board member of 20 prestigious (Indexed in SCI/SCIE/Scopus) international journals with high impact factor, i.e., *Applied Acoustics, Measurement, JESTEC, AJSE, SV-JME* and *LAJSS*. In addition, he has published 100+ research articles in journals, book chapters and conferences. He has authored/co-authored cum edited 22 books on Mechanical and Materials Engineering. He has published two patents. He is associated with International Conferences as Invited Speaker/Advisory Board/ Review Board member/Program Committee Member. He has delivered many invited talks in webinars, FDP and Workshops. He has been awarded as Best Teacher for excellence in academics and research. He has successfully guided 12 BTech, MTech and PhD theses. In administration, he is working as a coordinator for AICTE, E.O.A., Nodal officer for PMKVY-TI Scheme (Government of India) and internal coordinator for CDTP scheme (Government of Uttar Pradesh). He is currently involved in the research area of AI & ML in Mechanical Engineering, Advanced Materials & Manufacturing Techniques, Building Efficiency, Renewable Energy Harvesting, Heavy Vehicle Dynamics and Sustainable Transportation.

xviii About the Editors

Varun Pratap Singh received a PhD (Mechanical Engineering) in the area of Solar Energy. He is currently working as an Assistant Professor in the Department of Mechanical Engineering and Head-Solar Energy Centre at the College of Engineering Roorkee, Roorkee, India, since August 2011. He has more than 11 years of research and academic experience in mechanical engineering and renewable energy. He has published three patents and five research articles in SCI journals and conferences. He has completed five projects as Principal Investigator (PI) and two projects as Co-PI with a total research grant of more than twenty lakhs INR for the government as well as Industry partners. He has delivered many invited talks in webinars, FDP and Workshops and attended 15+ MOOCs and 15+ FDPs from various national bodies like IITs, AICTE and international bodies like Delft University of Technology-Netherlands, Tsinghua University-China, National Chiao Tung University-Taiwan and The International Monetary Fund (IMF). He has been awarded an award for outstanding mentoring of students and a perfect faculty award for excellence in academics and research. He has successfully guided 16 BTech, MTech and PhD theses. In administration, he is working as head of the Centre of Excellence, COER. As an academician Renewable Energy, Solar-Thermal application, Automation and Special Propose Machining (SPM) are his interest areas of association. His ORCID ID is 0000-0002-2148-5604.

Chandan Swaroop Meena is presently working as a Scientist in the Building Energy Efficiency Department and an Assistant Professor at the Academy of Scientific & Innovative Research (AcSIR) at CSIR-Central Building Research Institute, Roorkee (Ministry of Science and Technology, Government of India). He has done his PhD in thermal engineering from IIT Roorkee, and MTech in thermal engineering from NIT Kurukshetra. He does research in renewable energy utilization techniques, geothermal systems, building energy efficiency, two-phase flow and heat transfer. He is a Life Member of Indian Society of Heat and Mass Transfer (ISHMT), Life Member of India Building Congress (IBC) and Life Member of the Institution of Engineers India (IEI). He has published 29 research articles in reputed Journals and International Conferences. He is associated with International Conferences as Invited Speaker/Advisory Board/Review Board member. He has delivered many invited talks in webinars, FDP and Workshops.

Nitesh Dutt is working as an Assistant Professor in the College of Engineering Roorkee, COER University, Roorkee, Uttarakhand, India. He has more than 7 years of teaching experience. He has done his Bachelors in Mechanical Engineering, Masters and PhD from IIT Roorkee. He has published more than 11 research articles in international journals and conferences. His main research areas are nuclear engineering, heat and mass transfer, thermodynamics, fluid mechanics, refrigeration and air conditioning, computational fluid dynamics (CFD).

List of Contributors

Anjali Agrawal
Energy Centre
Maulana Azad National Institute of
  Technology
Bhopal, MP, India

Hozaifa Ahmad
Department of Mechanical
  Engineering
Al-Falah University
Faridabad, Haryana, India

Pranav Charkha
Department of Mechanical
  Engineering
G H Raisoni Institute of
  Engineering & Business
  Management
Jalgaon, Maharashtra, India

Muralimohan Cheepu
Department of Materials System
  Engineering
Pukyong National University
Busan, Republic of Korea

Madhur Chauhan
Department of Electrical
  Engineering
G H Raisoni Institute of
  Engineering & Business
  Management
Jalgaon, Maharashtra, India

Ashok Kumar Dewangan
Department of Mechanical
  Engineering
National Institute of Technology Delhi
Delhi, India

Ankit Dev
Department of Mechanical &
  Industrial Engineering
IIT Roorkee
Roorkee, Uttarakhand, India

Gaurav Dwivedi
Energy Centre
Maulana Azad National Institute of
  Technology
Bhopal, MP, India

Nitesh Dutt
College of Engineering Roorkee
COER University
Roorkee, Uttarakhand, India

Tauseef Hassan
Centre for Nanoscience and
  Nanotechnology
Jamia Millia Islamia
New Delhi, India

Md Hassaan
Department of Mechanical
  Engineering
Al-Falah University
Faridabad, Haryana, India

List of Contributors

S. Mojahid Ul Islam
Department of Mechanical
  Engineering
Al-Falah University
Faridabad, Haryana, India

D.B. Jani
Government Engineering College
Gujarat Technological University
  (GTU)
Dahod, Gujarat, India

Ashish Karn
Department of Mechanical
  Engineering
School of Engineering
University of Petroleum and Energy
  Studies
Dehradun, Uttarakhand, India

Sujeet Kesharvani
Energy Centre
Maulana Azad National Institute of
  Technology
Bhopal, MP, India

Osama Khan
Department of Mechanical
  Engineering
Jamia Millia Islamia
New Delhi, India

Arijit Kundu
Department of Mechanical
  Engineering
Jalpaiguri Government Engineering
  College
West Bengal, India

Ashwani Kumar
Technical Education Department
Kanpur, Uttar Pradesh, India

Vivek Kumar
Department of Mechanical
  Engineering
Netaji Subhas University of
  Technology
Dwarka, New Delhi, India

Mukesh Kumar
Mechanical Engineering
  Department
Malaviya National Institute of
  Technology
Jaipur, Rajasthan, India

Ravi Kumar
Department of Mechanical &
  Industrial Engineering
IIT Roorkee
Roorkee, Uttarakhand, India

Aditya Kumar
Department of Energy and
  Environment
NIT Tiruchirappalli
Tamil Nadu, India

Rajesh Kumar
Department of Mechanical
  Engineering
Dayalbagh Educational Institute
Agra, UP, India

Chandan Swaroop Meena
CSIR-Central Building Research
  Institute
Roorkee, Uttarakhand, India

Syed Quadir Moinuddin
Department of Mechatronics
  Engineering
ICFAI Foundation for Higher
  Education
Hyderabad, Telangana, India

Mohammad Ashad Ghani Nasim
Department of Mechanical
  Engineering
Al-Falah University
Faridabad, Haryana, India

Pragya Nema
Department of Electrical
  Engineering
Oriental University
Indore, Madhya Pradesh, India

List of Contributors

Neerja
PG Department of Physics and
 Electronics
DAV College
Amritsar, Punjab, India

Bipasa Patra
Department of Electrical
 Engineering
G H Raisoni Institute of
 Engineering & Business
 Management
Jalgaon, Maharashtra, India

Mohd Parvez
Department of Mechanical
 Engineering
Al-Falah University
Faridabad, Haryana, India

Md Adib Ur Rahman
Department of Mechanical
 Engineering
Al-Falah University
Faridabad, Haryana, India

Amit Sarin
Department of Physical Sciences
I K Gujral Punjab Technical University
Amritsar Campus, Punjab, India

Sanjeev K. Sajjan
Department of Mechanical
 Engineering, VCE
Warangal, Telangana State, India

R. P. Saini
Department of Mechanical &
 Industrial Engineering
IIT Roorkee
Roorkee, Uttarakhand, India

Naveen Sharma
Department of Mechanical
 Engineering
Netaji Subhas University of
 Technology
New Delhi, India

Noushad Shaik
Department of Advance Mechanical
 Engineering
University of Leicester
Leicester, United Kingdom

Varun Pratap Singh
Department of Mechanical
 Engineering
School of Engineering
University of Petroleum and Energy
 Studies
Dehradun, Uttarakhand, India

Meetu Singh
Department of Applied Sciences
I K Gujral Punjab Technical
 University
Kapurthala, Punjab, India

T. Sivasakthivel
Department of Mechanical
 Engineering
Global College of Engineering and
 Technology
Partnership Institute of the
 University of the West of
 England (UWE Bristol)
Muscat, Oman

Saurav Sikarwar
Department of Mechanical
 Engineering
Dayalbagh Educational Institute
Agra, UP, India

Deeptam Trivedi
Energy Centre
Maulana Azad National Institute of
 Technology
Bhopal, MP, India

Rahul Tarodiya
Department of Mechanical
 Engineering
Visvesvaraya National Institute of
 Technology (VNIT)
Nagpur, Maharashtra, India

Mrityunjai Verma
Department of Mechanical
  Engineering
School of Engineering, University of
  Petroleum and Energy Studies
Dehradun, Uttarakhand, India

Vikas Verma
Department of Energy
Tezpur University
Assam, India

Gulam Hasnain Warsi
Department of Mechanical
  Engineering
Jamia Millia Islamia
New Delhi, India

Ashok Yadav
Department of Mechanical
  Engineering
Dayalbagh Educational Institute
Agra, UP, India

Chapter 1

Introduction to Thermal Energy Resources and Their Smart Applications

Arijit Kundu, Ashwani Kumar, Nitesh Dutt, Chandan Swaroop Meena, and Varun Pratap Singh

CONTENTS

1.1 Introduction to Energy Resources: Basic Concepts 1
1.2 Energy Consumption: Brief History 2
1.3 Traditional Sources and Applications 4
1.4 Smart Applications of Thermal Energy Resources 7
References 13

1.1 INTRODUCTION TO ENERGY RESOURCES: BASIC CONCEPTS

Thermal energy has been utilized as the most fundamental type of energy for fundamental tasks including heating, cooking and boiling water from the beginning by the human species. The principal sources of thermal energy to be exploited in the beginning, in most cases, were fire for general combustion processes; and to continue it for a larger time period, we need coal, gas or oil, most of those are inexpensive, economic, but limited and environmentally filthy and risky. Later several additional natural and man-made sources, such as electrical heating, solar energy, geothermal energy, nuclear heat from the fusion and fission process, etc., have come into existence over time [1]. But these are unproven for large-scale use and obviously expensive; some are tremendously unsafe and hazardous to the environment also.

The first source of energy for humans was food, and that made it possible for them to live and procreate naturally. Other sources of energy recovered from thermal energy directly, or can be transformed into other types including mechanical, electrical and chemical energy, became more and more crucial to the advancement of humanity and expanded on this fundamental necessity. As hunter-gatherers, humans were able to survive on their own strength, skills and fire, as well as with their intelligence and knowledge. However, once people began to engage in established agriculture, they required new sources of energy to effectively cultivate the land

DOI: 10.1201/9781003395768-1

2 Thermal Energy Systems

and refine grain. For mobility, which is essential to modern civilization, they also required energy. Early means of transport were the backs of human slaves and animals. Then, sea transport was made possible by using sails to harness wind energy. Coal combustion-powered steam engines and fossil fuels brought the cars later [2]. After significant growth in civilization, the application of thermal energy has been evoked for the generation of direct or indirect electricity, in addition to the direct use of heat for countless household, industrial and manufacturing operations. Significant progress has been achieved in converting thermal energy into electrical energy as our reliance on the world of electricity grows. The combined cycle and co-generation technologies have significantly improved the thermal efficiency of power generation [1]. Thereafter, when mankind started visualizing the limited amount of capacity of conventional energy resources, renewable energies were being derived from natural resources that can be replenished, or replaced, through natural processes. Solar, wind, water, biomass and geothermal energy are examples of renewable resources that can never run out. The Sun continues to glow, the winds will blow, the rain will fall and the Earth continues to release and grasp heat no matter how much we utilize them.

1.2 ENERGY CONSUMPTION: BRIEF HISTORY

Wood was the primary fuel from the time that primigenius hominins gathered around roaring fires through the recent past. Up until 400 years ago, humans relied solely on renewable energy sources: they heated their homes and cooked their meals by burning wood, and they used to trek by using animal traction.

Gradually, humans created machines over time to perform tasks for them. Wind and water power were harnessed by machines such as windmills and waterwheels to power activities such as grinding grain. Coal mining sparked the invention of steam engines in the seventeenth century. With the onset of the Industrial Revolution in the late 1700s – a significant turning point in human history – that started to alter. The revolution ushered in the shift from manual labour to manufacturing based on machines. It began in England and swiftly expanded to the rest of the world, as well as to Europe and the United States. In the nineteenth century, oil, gas and nuclear power were all harnessed. Many of these sources have issues with pricing and accessibility.

James Watt (1736–1819), a Scottish engineer, built an effective steam engine in the late 1700s based on Thomas Savery's previous seventeenth-century design. In order to generate steam to power machinery, coal had to be burned. Over time, factories, ironworks, water pumps, steamboats and train locomotives were all powered by coal-burning steam engines. When engineers discovered that falling water might produce electricity in hydroelectric dams in the 1880s, water power saw a revival; the primary source of electricity was still coal-fired power stations, though. Gasoline began to gain popularity as an energy source when mass production of automobiles started in the early 1900s. Additionally, it has been discovered that natural gas is an excellent fuel for producing electricity, heating and cooking. The usage of petroleum and natural gas doubled in around 35 years. As this was happening, nuclear power entered the picture. The massive energy released by uranium atom splitting was put to use creating electricity.

By the end of the 1970s, the world's consumption of energy resources had become a significant issue due to unanimous population growth, industrial development and a corresponding increase in the need for coal for electricity and petroleum for transportation fuel. Due to the hazardous effect and limited supply of fossil fuels within the boundary of any country, either industrialized or underdeveloped, it prompted a search for clean, renewable and easily accessible alternative energy sources.

In general, the different sources of thermal energy comprise: (1) the combustion of fossil fuels (coal, gasoline, diesel, natural gas, heavy oil, etc.); (2) solar heat (active and passive); (3) nuclear heat fission; (4) geothermal and other renewable heat; (5) biomass and waste heat; (6) various electrical heating systems. Waste heat is arguably the least known but most significant source of heat because it is produced from all other sources of heat and is growing quickly as we use the other thermal energy sources more and more. By assisting them in feeding themselves, staying warm and other basic needs, humans can improve their welfare through the expenditure of energy. The distinction between primary and final energies is made. Between generation and consumption, primary energy has not undergone any conversion. Crude oil is a basic energy source, but refined fuels such as gasoline and diesel are secondary energies. Nuclear energy is classified as secondary, while electricity generated by hydropower or photovoltaic panels is classified as primary. Wood is used to create charcoal, a secondary energy source (primary energy). Oil, coal, natural gas, hydropower, wood, solar and wind energy all fall under this category. Customers' ultimate energy delivery can be used to satisfy their energy or non-energy needs. As shown in Figure 1.1, this distinction between primary and secondary energy might have an impact on how we assess and compare various energy sources. On a worldwide scale, nuclear energy and hydropower generate almost the same amount of electricity for the consumer. Despite the fact that the output from these two sources is equal, statistics reveal that nuclear energy actually produces four times more primary energy than hydropower (Figure 1.1). The electricity is produced with an efficiency that is very close to 100% using hydropower, which is a primary source. On the other hand, nuclear power's electricity, which is the energy released during uranium fission, is not considered a fundamental energy [3]. Although consumers utilize the same amount of energy from both sources, nuclear energy produces four

4 Thermal Energy Systems

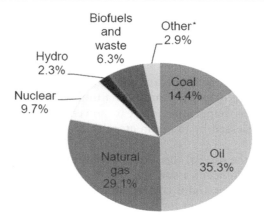

Figure 1.1 World consumption of primary energy in 2020. *Includes geothermal, solar, wind, ocean, heat and other sources.

Source: IEA World Energy Outlook 2020.

times more primary energy than hydropower due to the efficiency of today's power plants, which is roughly 25%.

Until about 1980, the World had a major reliance on macrothermal energy processes. The economy of scale was applied in huge, centralized enterprises, and more work was done to provide heat and power. Using thermoelectric, thermo-photovoltaic, thermo-ionic, thermo-galvanic and thermo-acoustic processes, heat was directly transformed into electricity as well as indirectly (through thermodynamic cycles and heat engines) [1]. The three fundamental laws of thermodynamics, which assume continuum mechanics and state that thermal energy depends on the basic physical properties of the materials, such as thermal conductivity, specific heat, melting and freezing points, and heats of vaporization, are responsible for controlling these processes. These characteristics are universal to all materials and do not depend on size.

1.3 TRADITIONAL SOURCES AND APPLICATIONS

From personal comfort to the progressive agenda for modernization and environmental preservation, thermal energy, the most fundamental form of energy, is required in many parts of society. The oldest known natural and man-made process is probably the combustion of combustible fossils and biofuels into thermal energy. In actuality, this energy conversion in the combustion process converts chemical energy to thermal energy. External combustion engines, such as early steam engines, were massive machines that took up entire buildings and were typically used to pump water out of flooded mines. They had a single cylinder and piston connected to a large beam, and they swayed back and forth. The injection of steam into the cylinder drove the piston to rise and the beam to lower. The steam was then cooled, a partial vacuum was created, and the beam tipped back in the other direction as a result of water being sprayed into the cylinder. Despite being a tremendous scientific advancement, steam engines were too large, slow and inefficient to power trains and industrial machinery. James Watt created a steam engine in the 1760s that was more manageable, powerful and affordable. Both fixed steam engines that could be used in factories and portable, movable steam engines that could power steam trains were created by Watt [4].

The development of gasoline-powered internal combustion engines by Lenoir (1822–1900), Marcus (1831–1898) and Otto (1832–1891) in the middle of the nineteenth century required cylinders, pistons and a spinning crankshaft in order to harness their power. Later in the nineteenth century, Diesel (1858–1913) discovered he could build internal combustion engines that were much more powerful and could run on a variety of different fuels [5]. Diesel fuel burns spontaneously and releases the contained heat energy because diesel engines compress the fuel much more than gasoline engines do. The best conversion efficiency now available for combined cycle steam and gas turbine systems is only about 60%, despite the fact that James Watt's steam engine has advanced significantly over the past 250 years. Different heat engines that have been proposed or are now in use have varied levels of efficiency. For instance, the efficiency of the planned ocean thermal energy conversion (OTEC) ocean power is about 3% [6] (97% waste heat using low-quality heat), while it is roughly 25% [7] for the bulk of car gasoline engines. A steam-cooled combined cycle gas turbine has an efficiency of about 60%, whereas supercritical coal-fired power plants like the Avedøre Power Station and many others have an efficiency of about 49% [8]. One of the primary objectives of the heat recovery process is to increase these efficiency numbers using novel and innovative ways. A modern, high-performance automobile engine produces more than 75 kW/L.

Solar energy, which is created by the sporadic fusion process, has recently become the most sought-after renewable energy source. The bulk of solar energy must first be converted to thermal energy before being used, despite the fact that photovoltaic technology allows for the direct conversion of solar radiation to electricity. Its utilization determines how solar energy is recovered. Solar energy can be collected centrally, as in concentrated solar power plants, or distributed, as in the HVAC (Heating, Ventilation and Air Conditioning) and water heating systems of individual homes. Solar heating has the ability to create both high and low temperatures. Thermal energy storage is required for the reliable use of solar energy due to its erratic nature.

Nuclear energy's fission, which generates a substantial amount of nuclear heat, is another man-made process. Electrical heating refers to a number of different ways in which electrical energy is converted into thermal energy.

Geothermal energy can be generated by both deeper and closer-to-the-surface hydrothermal systems. Geothermal energy can also be extracted from deep, hot and dry rock by using enhanced geothermal systems. Geothermal energy can be produced by both open and closed mines. Additionally, a portion of the energy derived from oil and gas wells that are already in place as well as geo-pressured zones can be used to produce geothermal heat in hybrid forms. Finally, geothermal energy can be produced by heated magma. Each method generates a different quantity of heat and therefore a different approach to recovery and usage.

Thermal energy from direct electrical heating can be produced using seven different techniques: joule heating, UV and IR heating, induction heating, microwave heating (or dielectric heating), electric-arc heating, plasma heating and power beam heating (such as electron beam and laser beam heating). Each functions in accordance with a separate electrical heating theory and employs a different electromagnetic wave and electricity spectrum. Each gets retrieved in a unique way and is put to use in a unique way. These recovery techniques are continuously improved in terms of their thermal efficiency and application scope. Recently, electrical energy has been used more frequently at these scales as a result of improvements in micro and nano technologies.

Due to the fact that most industrial waste heat is wasted and released straight into the ground, industrial operations have a significant potential for waste heat recovery (WHR). The difficulties in recovering waste heat on a technical and financial level are the main cause. Waste heat energy from various sources is depicted in Figure 1.2 [9].

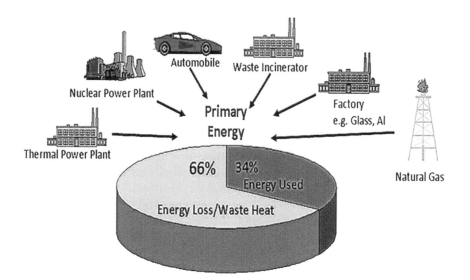

Figure 1.2 Waste heat energy from various sources [9].

To increase the effectiveness of enterprises and processes, thermal energy from various industries must be collected and stored. Industries that use a lot of energy, such as those that manufacture food and glass, cement, steel, oil and gas, are in the spotlight because they use a lot of energy and emit a lot of waste heat into the atmosphere. A latent heat thermal energy storage system and a waste heat releasing system can be coupled to estimate WHR with accuracy and precision. Depending on the characteristics of the thermal energy storage material, the size of the processing industry, the environment, etc., the percentage of waste heat that may be recovered can range from 45% to 85% [10].

1.4 SMART APPLICATIONS OF THERMAL ENERGY RESOURCES

The importance of smaller and more scattered thermal energy applications increased as a result of a revolution that occurred in the latter decades of the previous century. It was realized that centralized large-scale macrothermal processes cannot be properly optimized to suit all needs without an understanding of microthermal processes. The development of microelectronics and microlevel manufacturing processes called for a deeper understanding of thermal phenomena at lower sizes. As long as the basic principles governing thermal processes remained in place, microprocesses were initially just simple extensions of macroprocesses. Classical thermodynamics has been replaced by the statistical thermodynamics theory. The knowledge and modelling of microthermal processes have improved. For thermal processes, the top-down approach to optimization was still in use. The fundamental thermal properties of materials at the microscale were linked to the macro properties using probability functions.

Around the turn of the century, a brand-new paradigm of nano thermal processes began to take shape. A new understanding of thermal processes was required because none of the fundamental laws of the macro world applied to the world of nano thermal processes. This required a new bottom-up approach to thermal processes and allowed for the production of unique materials and devices that were not achievable using the top-down technique used in the macro world. As a result, the processing, control and recovery of thermal energy in the nano-world opened up new possibilities that were impractical in the macro world.

> "…an approach in which smart electricity, thermal, and gas grids are combined with storage technologies and coordinated to identify synergies between them in order to achieve an optimal solution for each individual sector as well as for the overall energy system [11]" is how smart energy systems are defined.

8 Thermal Energy Systems

Renewable energy sources such as solar and wind power will be a key component of the future energy grid. These resources don't have a lot of energy that is already stored; instead, energy from the wind, sun, waves and tides must be quickly captured and used. One of the main technological challenges that energy systems will face in the future is this. How will the future energy system, which will be based on renewable energy, function without the flexibility currently offered by significant amounts of stored energy in fossil fuels, while both delivering affordable electricity and making sustainable use of the resources at hand? The answer is in developing new energy system flexibility measures that are both cost-effective and effectively make use of renewable energy sources. Such a system is referred to as smart energy.

A smart energy system is made up of new infrastructures and technologies that produce new types of flexibility, mainly during the energy system's "conversion" stage. This is accomplished by moving away from the current energy systems' straightforward linear approach (i.e., fuel to conversion to end-use) and toward a more integrated one. To put it simply, this entails merging the thermal, transportation and electricity sectors in order to make up for the limited flexibility of renewable energy sources such as wind and solar. Figure 1.3 shows the smart energy system, which employs innovations as follows:

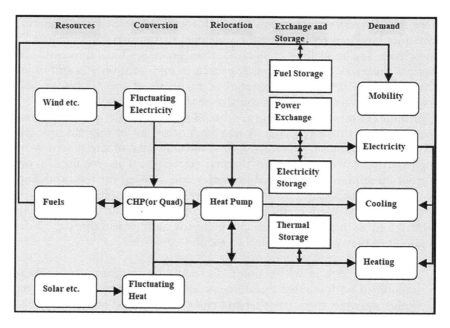

Figure 1.3 Example of a smart energy system based on the infrastructures [11].

1. **Smart Electricity Grids** link intermittent renewable energy sources such as wind and solar power with flexible electrical uses like heat pumps and electric automobiles.
2. Connecting the electricity and heating sectors with **smart thermal grids** (district heating and cooling). This makes it possible to recycle heat losses in the energy system and use thermal storage to add to the system's flexibility.
3. **Smart gas grids** link the transportation, heating and energy industries. This makes it possible to use gas storage to add more flexibility. Storages for liquid fuel can also be used if the gas is processed to become one.

Future energy systems must contend with the challenge of meeting transportation demands without going over the sustainable biomass potential in addition to the storage issue already mentioned. Direct usage of electricity, as shown in Figure 1.4, is the optimum option from a system perspective. Direct usage of power, however, is insufficient to meet all transportation needs. Biomass will have to be used to fill some of the gaps, but the amount of biomass that can be used for energy is constrained by the need for food and materials as well as biodiversity. Furthermore, they are so constrained that it is difficult to imagine how biomass alone could meet the current energy needs of the transportation industry. The claim is that power must be used directly in batteries for transportation in order to supplement some biomass that must be transformed into gas or liquid fuel. Furthermore, biomass in the form of gas helps future power plants is more adaptable and efficient.

Biomass is not the only relevant factor in gas generation. When paired with methods for producing gas from biomass, such as fermentation, gasification and hydrogenation, such conversion technologies may have significant

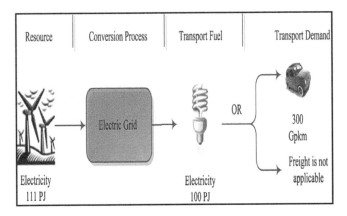

Figure 1.4 Direct use of electricity for transportation using battery storage [12].

10 Thermal Energy Systems

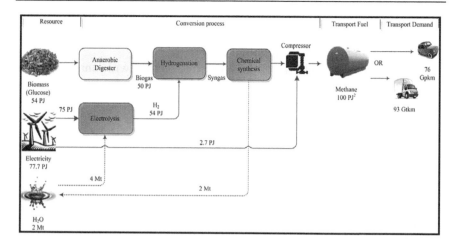

Figure 1.5 Illustration of using power-to-gas for hydrogenating synthetic gas derived from biomass [12].

synergies. One will need to increase the synthetic gas and liquid fuel from the biomass by hydrogenation, or in the form of creating hydrogen on electrolysers, in order to produce enough gas and liquid fuel for the entire renewable energy system within the biomass resources available. For various types of biomass resources and end users, there are many paths. One such approach is shown in Figure 1.5.

Future renewable energy systems are often designed using a combination of residual resources such as trash and biomass and variable renewable energy sources such as wind, geothermal and solar power. Future renewable energy systems that may be implemented realistically must incorporate significant components of energy conservation and energy efficiency measures in order to reduce the strain on biomass resources and the investments in renewable energy [12]. The conversion of solid biomass fractions into various liquid and gaseous biofuels for use as transportation fuel, among other things, maybe one of the industrial processes used in the future.

Making the appropriate decisions that would result in a better level of energy self-sufficiency presents various difficulties. Technical, institutional, legal and socio-cultural factors are among them. For any strategy to be successful, the plans should be built on locally accessible resources. Although expertise and technical information can be obtained abroad, the primary resources must be local. In this context, the region's relatively well-distributed renewable energies, particularly solar, biomass, wind and minor hydro potentials, would be extremely important as energy resources for those nations lacking access to traditional energy sources. As has been highlighted, rural settlements in underdeveloped nations are barely planned and are consequently scattered randomly without taking into account the possibility that some facilities may need to be shared. Every family fights to use the resources that are easily accessible while remaining independent. This way of life does not promote the creation of complex and effective methods that call for the coordinated effort of community members. People frequently still use outdated, ineffective approaches to meet their demands as a result, and as a result, efforts are typically focused on the fundamentals. As a result, the only energy needs of rural households are for lighting and cooking, as individual families lack the resources to obtain adequate energy for other uses. These characteristics unmistakably indicate that decentralized and renewable energy sources are the best energy options for rural communities. Due to the long distances that the experts would have to travel to reach isolated spots, maintenance and service expenses would also be too high for the utility companies. Due to the extremely low supply reliability, users would lose faith in this source of energy, be forced to turn to alternative sources of energy, and most likely would return to the old way of doing things. Since settlement patterns and quality of life differ from place to place, it is impossible to generalize such an occurrence trend to all rural areas [13]. Decentralized energy systems continue to be the secret to rural electrification, nevertheless. The best options for rural applications include widely available energy sources including solar, small hydro, wind and biomass, which also provide the finest decentralized supply methods.

For the identification of pioneer development and growth centres where they can be controlled by small rural units such as homes, villages or other structured community structures, it is crucial to first and foremost obtain reliable data on their distributions and potentials. The administration and ownership duties can also be delegated at these levels, with the local village chiefs and other leadership structures responsible for ensuring that the facilities are properly run and routinely maintained.

If the users value the services offered, the systems are more likely to continue operating this way. To provide first highly sought services and to show the feasibility of such resources, it is crucial to find suitable nucleus centres where community energy facilities might be developed. Afterwards, if there is a need and a suitable management structure, the services may be extended to the nearby villages. Schools, hospitals and formally established community centres could serve as such nuclei. Therefore, community involvement is a crucial requirement for the success of these decentralized systems. The energy supply provided by these facilities should be assessed every few years to determine their suitability and capacity to adapt to the evolving conditions, nevertheless, due to the growing population and rising energy demand [14,15].

The new arrangements must all work with the ongoing energy initiatives, such as the regulatory controls on independent power production. Effective coordination between the country's industrial energy needs and its rural energy needs is also more important than ever. The inability of a single supplier to accomplish the two energy supply goals without running the danger of neglecting one has hindered previous rural energy development strategies. For a power producer, the high concentration of industry in the main towns and their consistent, sizable payments to the utility companies are too alluring to contemplate the challenging management of the rural energy network. To be successful on both fronts, various but well-coordinated strategies must be implemented, and in this regard, co-generation should be a rural renewable energy effort that can connect well with the national energy supply network [16–20].

The majority of renewable energy technologies have advanced to the point where they may successfully fulfil their intended functions in developing nations. Unfortunately, research and development for these technologies were conducted outside of most developing regions, with little or no local input. The majority of them discovered that using low-level, conventional biomass-based technologies such as charcoal and wood stove advancements, which have had little impact on energy consumption patterns, was convenient and less expensive. While the usage of energy from co-generation plants was constrained by energy management legislation, attempts to cover biogas and gasification conversion processes have not been successful. This implies that extensive research will be required to lay the foundation for a suitable and successful response to processes of technology transfer and changes in energy policy. It will be necessary to import numerous renewable energy technologies, particularly photovoltaic technology and huge wind turbines. Simple technologies such as solar water heaters, solar cookers (both household and institutional), small hydro and wind turbines should be produced by local staff utilizing materials that are readily available in the area. All installations and maintenance of renewable energy systems should be done with the aid of locally created skills [21–23].

It is vital to instil an energy culture in people in order to do this so that society is aware of how to use and preserve energy from various sources [15]. In order to deal with all facets of renewable energy technologies at all levels, from craftsmanship to skilled applied scientists who can lead further developments and modifications appropriate for local conditions, it is also necessary to develop the critical mass of human resources. In this context, this term refers to the manpower, skills, knowledge and accumulated experience in dealing with all aspects of renewable energy technologies. These folks need to be placed in jobs that correspond to their educational backgrounds. In any development process, it is essential to place qualified employees in settings where they may successfully use their professional skills. The region's higher education institutions would need to improve their current renewable energy technology courses in order to increase the centre's capability to an effective productive level. The development of renewable energy technologies has made it unnecessary for the centre to waste resources on trying to reinvent the wheel. Instead, it should concentrate on resource identification and technology adaptation to local socioeconomic and resource accessibility conditions. The decision on this assignment should not be left just up to the centre's leadership because it is significant and vast. The operations must address the national energy development goals outlined in national development plans, which necessitates a sufficient annual budget allocation with funding disbursement based on the advancement of the goals. For obvious reasons, the demand for clean energy in metropolitan areas is much larger than it is in rural ones, and is typically higher than what can be accommodated by current infrastructure. Because of the high population density and the extensive economic and industrial activity, garbage is also produced in great quantities in metropolitan areas. Almost all municipal administrations in the area have struggled greatly with managing this garbage. In addition to being a net revenue sink, its collection and disposal are also extremely expensive. Waste may be turned into a resource that can help these municipalities make some money if it is managed properly. As a result, there would be greater demand and ability to upgrade sanitary facilities, which would lower the prevalence of diseases brought on by unhygienic living conditions. Less than 10% of the total population in the region directly benefits from these energy resources, though, in terms of accessibility. The remaining population, which is primarily rural, is required to locate and administer its own energy sources for reasons that go beyond accessibility. Poverty, temporary housing of poor quality, and the time-consuming preparation of regional specialty meals are also issues. Due to a number of limitations brought on by these circumstances, the people are forced to use the energy resources that are readily available and reasonably priced in their region, and as only biomass-based fuels, notably firewood, fall into this category, they are the obvious choice [24,25].

REFERENCES

1. Shah, Y.T., *Thermal energy: Sources, recovery, and applications*. Boca Raton, FL, USA: CRC Press, 2018.
2. Grimm, G., *Les nouvelles de l'environnement*. no. 57.
3. Heinrichs, A., *Environment at risk: Earth's energy resources*. Singapore: Marshall Cavendish Benchmark, 2011.
4. *External combustion engine*. Wikipedia. Retrieved from https://en.wikipedia.org/wiki/Externalcombustionengine, Last modified April 19, 2017.
5. *Internal combustion engine*. Wikipedia. Retrieved from https://en.wikipedia.org/wiki/Internalcombustionengine, Last modified April 22, 2017.
6. Dincer, I. and Rosen, M., *Thermal energy storage: Systems and applications*. New York: John Wiley & Sons, 2002.
7. Tari, A., *The specific heat of matter at low temperatures*. New York: World Scientific, 2003.
8. *Sensible heat*. Wikipedia. Retrieved from https://en.wikipedia.org/wiki/Sensibleheat, Last modified March 2, 2017.
9. Lindal, B., *Industrial and other applications of geothermal energy*, in: Armstead, H.C.H. (Ed.), *Geothermal energy: Review of research and development*. Paris, France: UNESCO, pp. 135–148, 1973.

10. Samuels, G., *Geopressure energy resource evaluation*. Oakridge, TN: DOE, Office of Energy Technology, Oakridge National Laboratory, p. 72, Available from NTIS, Springfield, VA, 1979; Schmidt, G.W., Am. Assoc. Petrol. Geol. Bull., 57, 321–337, 1973.

11. Lund, H., *Renewable energy systems: A smart energy systems approach to the choice and modeling of 100 % renewable solutions*, 2nd ed. Burlington, USA: Academic Press, 2014.

12. Lund, H., Andersen, A.N., Østergaard, P.A., Mathiesen, B.V., and Connolly, D., From electricity smart grids to smart energy systems – A market operation based approach and understanding, *Energy*, 42, 96–102, 2012.

13. Smale, T.H., *Cogeneration of heat and power: A market opportunity*, in: Twidell, J. (Ed.), *Energy for rural and island communities*. Oxford, UK: Pergamon Press, 1980.

14. Othieno, H., Alternative energy resources. *Journal of Energy Sources*, 14(4), 405–410, 1992.

15. Okken, P.A., Swart, R.J., and Swerver, S., *Climate and energy*. Dordrecht, The Netherlands: Kluwer Academic Publishers, 1989.

16. Singh, V.P., Jain, S., Karn, A., Dwivedi, G., Kumar, A., Mishra, S., Sharma, N.K., Bajaj, M., Zawbaa, H.M., and Kamel, S. Heat transfer and friction factor correlations development for double pass solar air heater artificially roughened with perforated multi-V ribs. *Case Studies in Thermal Engineering*, 39, 102461, 2022, ISSN 2214-157X, 10.1016/j.csite.2022.102461

17. Meena, C.S., Kumar, A., Roy, S., Cannavale, A., and Ghosh, A. Review on boiling heat transfer enhancement techniques. *Energies*, 15, 5759, 2022, doi: 10.3390/en15155759

18. Singh, V.P., Jain, S., Karn, A., Kumar, A., Dwivedi, G., Meena, C.S., and Cozzolino, R. Mathematical modeling of efficiency evaluation of double-pass parallel flow solar air heater. *Sustainability*, 14, 10535, 2022. doi: 10.3390/su141710535

19. Meena, C.S., Kumar, A., Jain, S., Rehman, A.U., Mishra, S., Sharma, N.K., Bajaj, M., Shafiq, M., and Eldin, E.T. Innovation in green building sector for sustainable future. *Energies*, 15, 6631, 2022. doi: 10.3390/en15186631

20. Singh, V.P., Jain, S., Karn, A., Kumar, A., Dwivedi, G., Meena, C.S., Dutt, N., and Ghosh, A. Recent developments and advancements in solar air heaters: A detailed review. *Sustainability*, 14, 12149, 2022. doi: 10.3390/su141912149

21. Dutt, N., Binjola, A., Hedau, A.J., Kumar, A., Singh, V.P., and Meena, C.S., Comparison of CFD results of smooth air duct with experimental and available equations in literature. *International Journal of Energy Resources Applications (IJERA)*, 1(1), 40–47, 2022. doi: 10.56896/IJERA.2022.1.1.006

22. Singh, V.P., Jain, S., and Kumar, A., Establishment of correlations for the thermo-hydraulic parameters due to perforation in a multi-V rib roughened single pass solar air heater. *Experimental Heat Transfer*, 2022. doi: 10.1080/08916152.2022.2064940

23. Kushwaha, P.K., Sharma, N.K., Kumar, A., and Meena, C.S. Recent advancements in augmentation of solar water heaters using nanocomposites with PCM: Past, present & future, *Buildings*, 2023; 13:79. 10.3390/buildings13010079

24. Singh, V.P., Jain, S., Karn, A., Dwivedi, G., Alam, T., and Kumar, A. "Experimental assessment of variation in open area ratio on thermo-hydraulic performance of parallel flow solar air heater". *Arabian Journal for Science and Engineering* 2022, 10.1007/s13369-022-07525-7

25. Meena, C.S., Prajapati, A.N., Kumar, A., and Kumar, M., Utilization of solar energy for water heating application to improve building energy efficiency: An experimental study. *Buildings*, 2022; 12: 2166. 10.3390/buildings12122166

Chapter 2

Thermal Energy Storage
Opportunities, Challenges and Future Scope

Ashok K. Dewangan, Syed Quadir Moinuddin, Muralimohan Cheepu, Sanjeev K. Sajjan, and Ashwani Kumar

CONTENTS

2.1 Introduction 17
2.2 Challenges and Characteristics of TES Systems 20
2.3 Thermal Energy Storage Technologies 20
    2.3.1 Sensible Heat Storage 21
        2.3.1.1 SHS Material 23
    2.3.2 Latent Heat Storage 23
    2.3.3 Chemical Heat Storage 24
2.4 Performance Parameters for TES 25
2.5 Solar System TES 25
2.6 Conclusions and Future Directions 26
References 27

2.1 INTRODUCTION

The current scenario demands a reduction of carbon emissions and an increase in the storage of resources and their energies due to global warming and the continuous reduction of conventional energy resources, respectively. This has led to the search for alternative and sustainable non-conventional energy resources and storage. Energy storage (ES) is defined as storing energy in one form and can be utilized in the same/another form whenever required for a specific application. The energy stored can be in different forms, such as mechanical, electrical, chemical, thermal, magnetic, etc., and its classification has been represented in Figure 2.1. However, there are different appropriate techniques to store energy in systems and the device that stores energy is known as an accumulator (Gil and Medrano 2010). This chapter focuses on the thermal energy storage (TES) features such as challenges, technologies, materials and applications. Recently, the TES has gained attention and plays a vital role in various engineering applications of heating and cooling buildings, water coolers and industrial

DOI: 10.1201/9781003395768-2

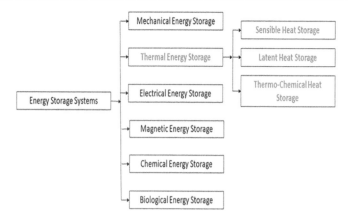

Figure 2.1 Classification of energy storage systems.

applications, etc. (Dincer 2002). TES is the amount of energy stocked thru heating or cooling action and utilized when needed.

Figure 2.2 depicts the TES process layout (Sadegi et al. 2022). Mostly TES bridges the gap between the energies (supply and demand) and also improves the efficiency and reliability of the system. The advantages of the TES system include economic efficiency, good reliability, reduced pollution, etc. TES systems are commonly classified based on energy storage mechanisms wherein sub-classified into three types (i) sensible heat storage (SHS), (ii) latent heat storage (LHS) and (iii) chemical heat storage (CHS) as shown in Figure 2.1 (Hasnain 1998, Dincer 2002, Gil and Medrano 2010,

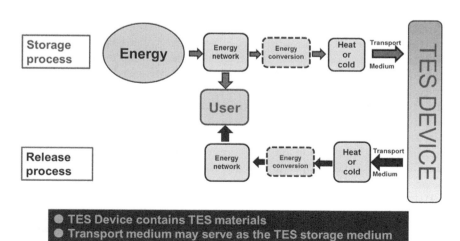

Figure 2.2 TES process layout (Sadeghi 2022).

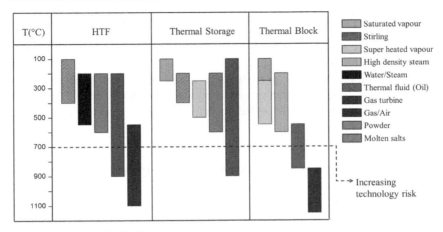

HTF: Heat transfer fluid

Figure 2.3 Classification of TES systems based on temperature ranges (Zhang et al. 2016).

Zhang et al. 2016, Sarbu and Sebarchievici 2018). The other classification available in the literature were (a) based on temperature range and re-use technology, as shown in Figure 2.3, wherein the temperature ranges depend on the type of hot thermal fluids (HTF) selected and thereby, TES is a function of HTF. The traditional HTF involves molten salts, thermal fluids, water, steam, gas and particle suspensions (Zhang et al. 2016). In addition, TES can be classified as low-temperature TES and high-temperature TES, (b) based on storage concepts including active, passive and hybrid, as shown in Figure 2.4 (Kuravi et al. 2013). In the case of an active storage system (ASS), the HTF circulates through the system, whereas the passive storage system (PSS) does not circulate throughout the system. The charging and discharging done by forced convection in ASS and PSS, the HTF is responsible.

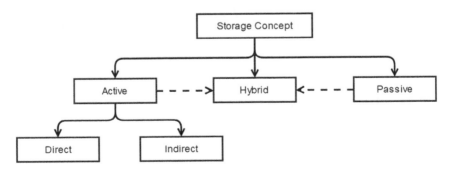

Figure 2.4 Classification of TES systems based on storage concept (Gil and Medrano 2010, Zhang et al. 2016).

The PSS uses phase change materials (PCM), concrete, etc., as a storage medium. Hybrid storage is a combination of active and passive that enhances storage systems. This chapter focuses on the challenges involved in TES technologies, i.e., sensible heat, latent heat and thermochemical, TES materials, performance parameters and applications in solar systems. The aforementioned topics are briefly discussed in subsequent sections that allow enlightening the opportunities and future research directions.

2.2 CHALLENGES AND CHARACTERISTICS OF TES SYSTEMS

The specific vital challenges that are associated with the TES systems [Sadeghi 2022] are as follows:

a. TES materials – the necessity to achieve high energy density, enhanced life span, wide temperature ranges and economic efficiency.
b. TES device – suitable advanced manufacturing process to generate TES devices, based on demand, the rate of control on charging/discharging process, reduced interfacial thermal resistance, compact design, low cost, enhanced life span and efficiency.
c. TES system – essential to integrating with energy networks to improve system reliability, performance and dynamics.

Further challenges associated with storage technologies and materials are discussed in the following sections. Before addressing the challenges, it is important to understand the characteristics of TES systems. The chemical reaction material has high capacity, power, efficiency and storage period than other TES systems, but its cost is high compared to other TES systems [Sabru et al. 2018]. The capacity defines the stored energy depending on the material and size. The power defines the rate of discharge of the system. Efficiency defines the ratio of output energy to input energy. The storage period defines the system's life in hours, weeks and months. The charging and discharging rate provide the time required to charge/discharge the system. The aforementioned variables are dependent on each other, and typical variable values are shown in Table 2.1. The selection of storage medium plays a significant part in TES systems.

2.3 THERMAL ENERGY STORAGE TECHNOLOGIES

To eliminate the gap between demand and energy supply, TES mostly uses technology to store heat energy. Sensible heat, latent heat and thermochemical heat storage are the leading TES technologies [Alva et al. 2017, 2018] and are discussed next.

Thermal Energy Storage   21

Table 2.1 Typical variable values for TES systems (Sabru and Sebarchievici, 2018)

Types of energy storage system	Capacity (kWh/t)	Power (kW)	Efficiency (%)	Storage period	Cost (€/kWh)
Sensible heat storage (SHS)	10–50	1–10,000	50–90	Days/months	0.1–10
PCM based	50–150	1–1,000	75–90	Hours/months	10–50
Chemical heat storage (CHS)	120–250	10–1,000	75–100	Hours/days	8–100

2.3.1 Sensible Heat Storage

The most straightforward technique to store thermal energy is heating or cooling a storage medium, as shown in Figure 2.5. Due to the effect of temperature gradient on the storage medium, the storage medium must have a sufficient amount of storage material, high specific heat capacity, sustained stability during the thermal cycle, and low cost and other thermal properties (Gil and Medrano 2010, Sarbu and Sebarchievici 2018). SHS is broadly classified based on storage media, such as liquid and solid storage media. The liquid storage media includes water, salt water, petroleum-based oils, molten salts, synthetic oils, etc. However, ignoring the detriments of storage system design and its stability, low density, cost increment, etc. Various SHS materials in liquid and solid mediums, such as sand, rock, concrete, granite, Aluminium, cast iron, water, octane, etc., have a temperature range from 20°C to 126°C with specific heat range from 800 to 2,400 J/kg.K, showing good storing heat capacity. The sensible liquid storage media uses a hot and cold fluid that density differences can distinguish. While charging and discharging, the hot fluids used are kept in the upper part of the storage, whereas cold fluids used are kept in the lower part of the storage system. In another mechanism, the fluid is supplied at the

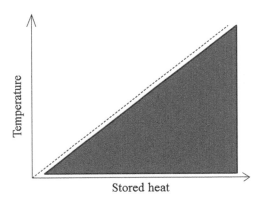

Figure 2.5 Sensible heat storage.

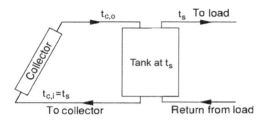

Figure 2.6 Typical liquid storage medium.

appropriate level with the temperature required such that the mixing of two fluids is avoided. Water is considered the best and most economical liquid storage medium of all liquid storage mediums. The amount of storage depends on the material's specific heat; the higher the specific heat, the more the storage. Water is an energy storage material that has applications over various temperatures, i.e., hot and cold forms. One such liquid storage medium application is the storage of a water tank that is economically efficient and readily availed. The typical water tank storage system shown in Figure 2.6 represents that the tank can be utilized as hot water storage due to solar heating and cold water storage through underground storage systems [Sabru et al. 2018].

The solid storage materials used for low and high temperatures involve sand-rock materials, fire bricks, concrete, ferroalloys, etc. It is noted that these materials work at temperatures ranging from 200°C to 1,200°C with good thermal conductivity and economic viability (Zhang et al. 2016). However, the major challenge is that their specific heats are low, ranging from 0.56 to 1.15, leading to a large storage unit. Unlike a liquid storage medium, the solid storage medium does not leak from the container. It is evident that the highest-density solid storage medium is cast iron, but availing the cast iron is more expensive than other solid storage mediums. Commonly used solid storage medium such as rock pile and pebble bed wherein the rock material is loosely packed, which transmit the heated fluid and stored by sending hot air and used again when required. In addition to material properties, the other parameters include size, shape, HTF, packing density, etc. The storage in the stack of rocks/bricks mostly applies to buildings wherein the rock size ranges from 1 to 5 cm. The significant advantages of SHS are that they are cheap, no risks involved in the usage of toxic materials. The stored heat depends on the specific heat and temperature difference.

The amount of heat stored in a storage material is given by the following equation:

$$Q_s = \int_{t_i}^{t_f} mC_p dt = mC_p(t_f - t_i) \qquad (2.1)$$

where Q_s is the heat stored [J]; m is the mass [kg]; C_p is the specific heat [J/(kg K)]; t_i is the initial temperature [°C]; and t_f is the final temperature [°C].

2.3.1.1 SHS Material

SHS materials are a set of materials which undergoes no change in phase over the temperature range (Gil and Medrano 2010). SHS materials, except liquid metals, are extensively used for low-cost materials. SHS is mainly used for high temperatures due to good thermal stability, better transport properties and heat transfer performance. The specific heat of SHS materials is smaller than LHS, which requires large volumes or quantities. The temperature of storage material decreases during the discharging process is a very crucial problem in SHS. The amount of thermal energy stored can be expressed as follows:

$$Q = \rho C_p V \Delta T \tag{2.2}$$

where Q is the amount of heat stored [J]; ρ is the density of the storage material [kg/L]; C_p is the specific heat over the temperature range of operation [J/(kg K)]; V is the volume of storage material used [L]; and ΔT is the temperature range of operation [°C].

In solid or liquid materials, storing thermal energy by sensible heat is easily possible. Castable and concrete ceramics, due to their economic cost and better thermal conductivity, are the most studied solid materials, and these materials are suitable for solid heat storage materials (Lovegrove et al. 1999). Materials such as mineral, synthetic and silicon oil, HITEC solar, nitrite, nitrate and carbonate salts, and liquid sodium are examples of solid storage (Herrmann and Kearney 2002).

2.3.2 Latent Heat Storage

LHS is an attractive approach that uses the stored energy when a substance changes from one phase to another at a constant temperature. LHS provides a high energy storage density and can store heat as latent heat of fusion. The storage capacity of an LHS with phase change material is

$$Q = m \int_{T_i}^{T_m} C_{ps} dT + m \Delta H_m + m \int_{T_m}^{T_h} C_{pl} dT \tag{2.3}$$

where T_m is the melting point temperature; C_{ps} and C_{pl} are the specific heat solid and liquid PCM; and ΔH_m is the phase change enthalpy.

LHS with PCM stores heat in large latent heat of fusion than SHS. This significant difference provides PCM with high energy density, reduces the surface area and the volume of the TES vessel and minimizes heat loss. The discharging process of LHS maintains the constant temperature and makes the adjacent space temperature stable compared to sensible heat. The low thermal conductivity of PCMs is the main drawback of LHS. PCMs such as inorganics salts show that an average number of freezing melting cycles degrade the performance of the materials.

PCMs, undergo solid–solid, liquid–gas and solid–liquid phase transformations. Few solid–solid PCMs have transition temperature, heat of fusion is suitable for thermal storage applications. Liquid–gas PCMs have a high heat of transformations and are thus not used for practical application due to the significant volume change during transformation. Solid–liquid PCMs are effectual as it stores a large amount of heat in a temperature range with no significant change in volume (Hasnain 1998). The PCM is stacked in a long thin tube container. Various applications of LHS with PCM are solar heating and cooling of the building, heat-pump systems, solar water heating and thermochemical storage (TCS) (Sarbu and Sebarchievici 2018).

Storage systems with solid–liquid transition are an alternative to sensible thermal storage (Gil and Medrano 2010). The PCM storage system operates between charging and discharging with a slight temperature difference in Figure 2.7 (Mehling and Cabeza 2008). It has a high energy density than sensible storage. Most PCMs have low thermal conductivity, leading to slow charging and discharging rates (Couto et al. 2008).

2.3.3 Chemical Heat Storage

Thermochemical heat uses reversible reactions to absorb, store and release energy between chemical reactants and solar heat. A thermochemical heat storage system has high volumetric storage density, lower volume and low temperature. Thermochemical heat storage system has some disadvantages: poor reactivity and the reaction of reversibility, coarse reaction and toxicity. The feasibility and durability of TES must be verified over a large

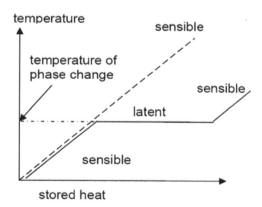

Figure 2.7 PCM phase change.

scale. CHS uses reversible reactions involving the absorption release of heat for TES (Gil and Medrano 2010). Typical advantages of this mechanism are storing high energy densities, pumping heat capability and indefinitely long storage duration at near ambient temperature. The generic chemical equation for TES is as follows (Shkatulov et al. 2015):

$$AB \underset{-Q}{\overset{+Q}{\rightleftarrows}} A + B \qquad (2.4)$$

The reaction depends on pressure and temperature. Heat is applied to material AB during the charging process, separating the two parts, A + B. The resulting reaction can easily be separated and stored until the discharge process. This two-part (A+B) is mixed with the appropriate pressure and temperature, resulting in energy release. CHS has advantages such as high TES density (Per-unit volume and mass), long duration of TES with nominal heat losses, etc. However, chemical thermal storage has a few disadvantages, such as when decomposition occurs during charging, resulting in lower porosity.

TES utilises heat energy to release higher energy for the efficient process than other technologies. For long-term heat storage, chemical storage material has theoretically zero heat loss [Guruprasad et al. 2018]. Many reactions are used as CHS materials, such as ammonia, methane, hydroxide, calcium carbonate, and iron carbonate. Liquid ammonia is a chemical material which dissociates in energy stored in the chemical reactor (Lovegrove et al. 1999). Ammonia is one of the most extensive chemical processes in industries. Ammonia has a strong pungent smell gas and is used for fertilizer and cleaning agent production (Lovegrove et al. 1999).

2.4 PERFORMANCE PARAMETERS FOR TES

SHS in the TES system offers a storage capacity ranging from 15 to 55 kWh/t and efficiencies between 52% and 90%, depending on the thermal insulation and specific heat storage medium (Ohanessian and Charters 1978). PCMs can provide more storage capacity and performance efficiency of up to 75–90%. The storage capacities of the solid–liquid phase and TES system can reach up to 250 kWh/t with a temperature of more than 300°C and 75–100% efficiency. PCM storage is more expensive and complex than the SHS system. Advanced heat transfer technology is used to enhance the performance of storage capacity (Sarbu and Sebarchievici 2018).

2.5 SOLAR SYSTEM TES

Solar energy systems have many applications in different sectors, such as building space heating, hot water supply, generating electricity and

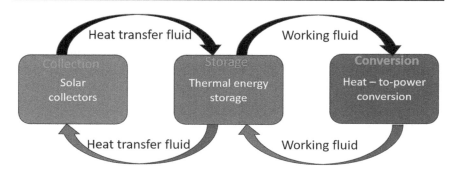

Figure 2.8 Solar thermal-driven power technologies.

concentrated solar power (CSP). Intermittence is one of the drawbacks of solar energy. To resolve this problem, use energy source backup or different hybridization sources of energy, i.e., gas and the electric grid for electricity. An alternate solution is to store heat during sunshine in a TES system and release it when required. Solar energy generates electricity that can be achieved by either a solar thermal-driven power cycle or solar photovoltaic (PV) (Wang et al. 2019). The driven power cycle in solar thermal collection, storage and conversion is shown in Figure 2.8. PV systems use batteries to store electricity to reduce the intermittence of solar energy. Compared to different energy storage technologies, the TES option is more economical than the battery choice for large scale (Chen et al. 2009). Generally used solar thermal-driven power technology is the CSP system, where high-temperature TES is present (Pantaleo et al. 2017, Song et al. 2019).

2.6 CONCLUSIONS AND FUTURE DIRECTIONS

This chapter reviewed the recent development in TES technologies, such as sensible, latent and thermochemical heat. Each of these technologies removes the energy difference through various working methods. The thermal energy system, its stability with material, thermal performance and techniques were analyzed. The typical storage materials used in solar energy systems are rock, sand and water due to their high specific heat, economical price, non-toxic and easy availability. The following conclusions have been drawn as follows:

- The thermal storage system must be efficient and cost-effective for the upcoming solar power development. The types of storage systems are SHS, LHS, and CHS. Solid SHS has recently been tested using concrete and castable ceramics. LHS is an advanced technology with a higher storage density at a constant temperature.

Various materials have been investigated and identified, but till now, the high-temperature PCM technology is unavailable. The three essential factors influencing the selection of PCMs in any application are melting temperature, latent heat of fusion and PCM thermo-physical issues. For solar power heat storage, CHS technology is less developed than latent heat. The commercially available sensible heat is TES systems, while PCM-based storage system is under development. SHS systems need a proper design to release thermal energy at a constant temperature.

- The prominent TES material is molten salt, with excellent thermal stability at a medium- and high-temperature power plant, low cost and viscosity. The temperature decrease during the discharging process is one of the drawbacks of SHS. Nowadays, LHS with PCM is a popularly used energy storage technology. It has a higher heat storage density at a constant temperature. Various storage materials such as organic, inorganic, eutectic and composite materials have been identified and analyzed. Several technologies, such as heat pipes, metal fins and other heat transfer technology, are explained to overcome low thermal conductivity. Chemical storage technology also has the potential for high storage density and long energy conservation. However, it is less popular than other solar energy storage methods.

The most common storage material for solar energy is studied in this chapter. In a solar power system, CSP is widely used for high-temperature TES. The low-temperature water PCMs materials are used and stored in tanks for a low-temperature solar power system. TES in the heating/cooling of the solar system is based on the lower temperature materials compared to CSP. The most widely used tanks for short-term storage are water tanks joined with solar thermal collectors for heat exchangers. Seasonal storage is large water tank thermal energy storage. The storage material is the construction material or part of the construction material. The development of TES material is crucial since the integration and design of the system are equally important for the stability of the solar energy system. Innovation in solar energy and thermal energy storage is highly desirable.

REFERENCES

Alva G., Liu L., Huang X., and Fang G. Thermal energy storage materials and systems for solar energy applications, *Renewable and Sustainable Energy Reviews*, 2017, 68, 693–706.
Alva G., Lin Y., and Fang G. An overview of thermal energy storage systems, *Energy*, 2018, 144, 341–378.
Chen H.S., Cong T.N., Yang W., Tan C.Q., Li Y.L., and Ding Y.L. Progress in electrical energy storage system: A critical review, *Progress in Natural Science*, 2009, 19, 291–312.

28 Thermal Energy Systems

Dincer I. On thermal energy storage systems and building applications, *Energy and Buildings*, 2002, 34, 377–388.

Do Couto Aktay K.S., Tamme R., and Müller-Steinhagen H. Thermal conductivity of high-temperature multicomponent materials with phase change, *International Journal of Thermophysics*, 2008, 29, 678–692.

Gil A. and Medrano M. State of the art on high-temperature thermal energy storage for power generation, *Renewable and Sustainable Energy Reviews*, 2010, 14, 31–55.

Hasnain S.M. On sustainable thermal energy storage technologies, Part 1: Heat storage materials and techniques, *Energy Conversion Management*, 1998, 39(11), 1127–1138.

Herrmann U. and Kearney D.W. Survey of thermal storage for parabolic trough power plants, *Journal of Solar Energy Engineering*, 2002, 124(2), 145–152.

Ioan Sarbu I.D. and Sebarchievici C. A comprehensive review of thermal energy storage, *Sustainability*, 2018, 10, 191.

Kuravi S., Trahan J., Goswami D.Y., Rahman M.M., and Stefanakos E.K. Thermal energy storage technologies and systems for concentrating solar power plants, *Program Energy Combustion Science*, 2013, 39, 285–319.

Lovegrove K., Luzzi A., and Kreetz H. A solar-driven ammonia-based thermo-chemical energy storage system, *Solar Energy*, 1999, 67, 309–316.

Mehling H. and Cabeza L.F. *Heat and cold storage with PCM*. Berlin, Heidelberg: Springer, 2008.

Ohanessian P. and Charters W.W.S. Thermal simulation of a passive solar house using a Trombe-Michel wall structure, *Solar Energy*, 1978, 20, 275–281.

Pantaleo A.M., Camporeale S.M., Miliozzi A., Russo V., Shah N., and Markides C.N. Novel hybrid CSP- biomass CHP for flexible generation: Thermo-economic analysis and profitability assessment, *Applied Energy*, 2017, 204, 994–1006.

Sadeghi G. Energy storage on demand: Thermal energy storage development, ma-terials, design, and integration challenges, *Energy Storage Materials*, 2022, 46, 192–222.

Shkatulov A. and Aristov Y. Modification of magnesium and calcium hydroxides with salts: 2059 An efficient way to advanced materials for storage of middle–temperature heat, *Energy*, 2015, 852060, 667–676.

Song J., Simpson M., Wang K., and Markides C.N. Thermodynamic assessment of combined supercritical CO_2 (sCO_2) and organic Rankine cycle (ORC) systems for concentrated solar power, *International Conference on Applied Energy 2019 (ICAE 2019)*, Västerås.

Wang K., Herrando M., Pantaleo A.M., and Markides C.N. Techno-economic as-sessments of hybrid photovoltaic-thermal vs. conventional solar energy sys-tems, *Case Studies in Heat and Power Provision to Sports Centres. Applied Energy*, 2019, 254, 113657.

Zhang H., Baeyens J., Cáceres G., Degrève J., and Lv Y. Thermal energy storage: Recent developments and practical aspects, *Progress in Energy and Combustion Science*, 2016, 53, 1–40.

Chapter 3

Introduction and Fundamentals of Solar Energy Collectors

Madhur Chauhan, Bipasa Patra, and Pranav Charkha

CONTENTS

3.1 Introduction 29
3.2 Material and Methods 30
3.3 Classification of Solar Collectors 32
    3.3.1 Flat Plate Collector (FPC) 33
    3.3.2 Compound Parabolic Collectors... 34
    3.3.3 Evacuated Tube Collectors (ETC) 35
3.4 Sun Tracking Collectors... 35
    3.4.1 Parabolic Through Collector 35
    3.4.2 Linear Fresnel Reflector... 36
    3.4.3 Parabolic Dish Reflector... 36
    3.4.4 Heliostat Field Collector 37
3.5 Discussion 38
References... 38

3.1 INTRODUCTION

With each passing year, the situation in energy becomes increasingly critical. The need for heat and electrical energy rises as the industry expands. It is commonly known that a significant portion of the total energy used is created through the burning of different fossil fuels, including solid, liquid and gaseous forms. The supplies of fossil fuels will run out in the near future since they are expensive, need a storage facility and pollute the atmosphere when burned [1]. The majority of energy conversion systems work with extremely controllable energy sources, allowing the system's energy consumption to be readily adjusted to get the desired outcome. For instance, a thermostat regulates the quantity of natural gas a home furnace uses to maintain a pleasant interior temperature despite the energy input being reduced due to system inefficiencies.

By supplying enough gas and maintaining a specific level of pressure in the gas pipelines, a gas company ensures that this system has access to

DOI: 10.1201/9781003395768-3

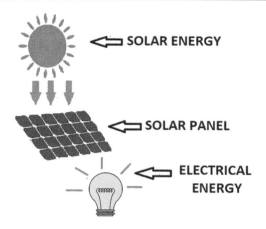

Figure 3.1 Solar energy conversion [4,5].

enough input. Switches regulate the quantity of input for end-use energy systems that use electricity, and the electric utility ensures that this input is enough. The quantity of coal fed to the boiler governs the amount of energy input for an intermediate energy conversion system such as a coal-fired electrical generator [1]. Solar energy is the available renewable alternative resource to satisfy the rising need for energy because of its accessibility, renewability, huge potential and environmental friendliness. However, the industrial applications of this free and green energy are severely constrained by enormous obstacles with the efficient collection and effective storage. Solar energy must be properly absorbed, delivered and stored since it is only used during the daytime. This means that in every solar-powered installation, sun collectors are the most crucial facet [2,3] (Figure 3.1).

3.2 MATERIAL AND METHODS

Four technical processes – electrical, chemical, thermal and mechanical – can utilize solar energy [6]. Through the chemical process of photosynthesis, which creates food and changes CO_2 into O_2, life on the Earth is possible. Several terrestrial uses and the powering of spacecraft involve electrical processes utilizing photovoltaic converters. Mechanical energy such as wind and water steams is another way to transform solar radiation [7]. A mechanical system known as a solar collector collects radiant sun energy and transforms it into usable heat energy [8]. Solar energy has been used to generate heat since ancient times. In the past, techniques for capturing and transporting solar energy were passive. In this system, heat is captured and transported via natural mechanisms such as irradiation, thermal conduction, heat transfer movement and the thermal material characteristics. Conversely, active solar heating techniques use pumps and

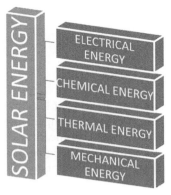

Figure 3.2 Utilization of solar energy [9].

fans to increase the speed at which heat and fluid are transferred (Figure 3.2).

Passive systems are those in which the thermal energy transfer occurs naturally, such as by natural convection, radiation and conduction [10]. Passive solar use, for instance, is when windows are positioned to maximize solar gains in order to satisfy winter heating needs, provide daylighting or both. One illustration of such a feature is the usage of insulation to reduce the warming or cooling capacity. Similarly, proper window placement or window shading could reduce solar gains and consequently summer cooling demands. Solar collectors and thermal batteries are examples of active solar systems, whereas south-facing windows and greenhouses are examples of passive solar systems. The conversion of the sun's radiant heat to heat is now the most popular and sophisticated way of solar energy conversion. The temperature and the quantity of this transformed energy are the two most crucial elements that must be recognized in order to properly match a conversion strategy to a particular task [9]. Temperatures that might be attained based on the concentration level are shown in Figure 3.3 [11]. Solar cookers, photoelectric (PV) systems, thermal energy, storage and concentrated sunlight power (CSP) are the main applications of solar energy. This topic introduces several solar collector systems, including solar parabolic plants, solar dishes and solar heliostat collectors, with a primary focus on solar

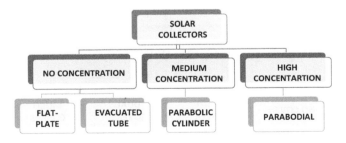

Figure 3.3 Classification of solar collectors [7].

collectors. We shall investigate, as an illustration, the energy and energy of a heliostat collection with such a chemical precipitation collector.

3.3 CLASSIFICATION OF SOLAR COLLECTORS

Most of the time, solar collectors are covered in order to avoid heat loss (the pool collector is an exception; when necessary, the obtainable temperature level is about 10–20°C above ambient temperature). The classification of solar collectors according to the quantity of covers is illustrated in Figure 3.4.

Heat losses must be minimized in order to increase the number of covers, but it should also be remembered that doing so will result in less solar revenue. For a system of N with identical materials, the analysis shown below is derived. The following equation [9] is used for calculating the transmittance for both the parallel and perpendicular components of polarization T_{rN} for different cover N.

$$T_{rN} = \frac{1}{2}\left\{\frac{1-r_n}{1+(2N-1)r_n} + \frac{1-r_p}{1+(2N-1)r_p}\right\}$$

where rN is the non-dimensional parallel component of unpolarized radiation; r_p is the uni-dimensional value of the perpendicular radiation component; N is the number of covers.

It is common knowledge that due to reflecting losses from the device surface, the maximum gain from a solar collector can only be achieved when sunlight strikes the device perpendicularly. Solar collectors are divided into three systems (Figure 3.5) [11].

The plant's performance will be optimized in the end by estimating the energy and energy waste of the loop [12]. In this sense, following the introduction of the collectors, the major categories will be stated clearly as the two most common types of solar collectors:

a. Static collectors
b. Sun tracking collectors

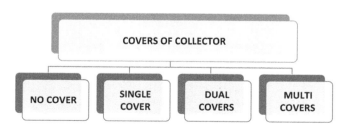

Figure 3.4 Number of covers of collector [9].

Fundamentals of Solar Energy Collectors 33

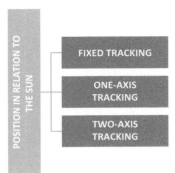

Figure 3.5 Equipment ratio of the collector [9].

Static collectors have a fixed location and do not move with the sun. Three types of collectors are described in this category:

i. Flat plate collector (FPC)
ii. Static compound parabolic collector (CPC)
iii. Evacuated tube collector (ETC)

3.3.1 Flat Plate Collector (FPC)

A flat plate collector works by using radiation that passes through a sheet of glass and is then placed on an absorber layer, which converts irradiation into thermal energy [13]. The heat is subsequently transmitted to a type of medium liquid, such as air in the tubes, water, or water with an anti-freeze component, to raise its temperature for immediate thermal use. To minimize conduction losses, the enclosure's interior and the underside of the absorber plate are both suitably insulated. The first order's opaque covering is used to stop solar reflectivity loss because it doesn't cross the long-wave photons. Additionally, the absorber surface's convection losses are decreased since the opaque covering restricts the amount of incoming air that can enter the space between it and the glass. Figure 3.6 shows a flat plate collector in perspective [14].

High-efficiency collectors can produce higher temperatures (water must be replaced with another heat transfer liquid because its boiling point is 100°C) (Figure 3.7).

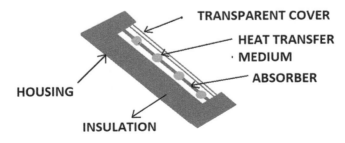

Figure 3.6 Flat plate collector [15].

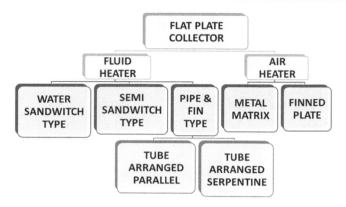

Figure 3.7 Flat plate collector [6].

3.3.2 Compound Parabolic Collectors

Winston created compound parabolic collectors [16]. Almost all of the light that is sent toward these collectors' mouths can be absorbed by them. These collectors have the capacity to concentrate a sizable part of diffuse radiation that strikes their pores without the sun being tracked [17]. Even while some varieties of compound parabolic collectors, which are similar to flat plate collectors, can track the sun's light, these collectors should be locked in a particular acceptance angle depending on their position. The accepted fact inclination for fixed CPC collectors installed in this mode is 47 degrees. From the summer to the cold weather solstices, the deflection of the sun is covered by this angle. Nevertheless, for creating a compound parabola. The gap should be kept as minimal as possible because it causes a loss in the reflector area and consequent performance loss. For flat receivers, this is more significant. Utilizing a tracking CPC is an option for applications requiring greater temperatures. Tracking is very erratic or intermittent when it is employed because the concentration ratio is normally low and there is a chance that radiation may be collected and concentrated by one or more reflections on the parabolic surfaces. CPCs can be created as two distinct units or as one unit with one opening and one receiver (Figure 3.8).

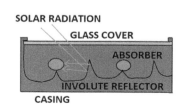

Figure 3.8 Compound parabolic collectors [18].

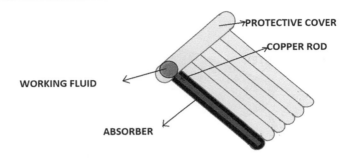

Figure 3.9 Evacuated tube collectors [19].

3.3.3 Evacuated Tube Collectors (ETC)

Sun tracking collectors, as the title suggests, follow the sun throughout the day to gather more energy than fixed collectors. Four primary sun tracking collector types will be discussed in this chapter. Sun tracking collectors, as the title suggests, follow the sun throughout the day to gather more energy than fixed collectors (Figure 3.9).

3.4 SUN TRACKING COLLECTORS

Four primary sun tracking collector types will be discussed in this chapter:

  i. Parabolic through collector
  ii. Linear Fresnel reflector
  iii. Parabolic dish reflector
  iv. Heliostat field collector

3.4.1 Parabolic Through Collector

A collection of concave mirrors called a parabolic trough focuses sunlight onto a receiver tube that is situated in the focus. These troughs can monitor the Sun around a single axis, usually, one that is oriented north-south for maximum effectiveness. This tube's fluid circulates through it while absorbing heat from the focused solar radiation. A linear Fresnel system has features in common with a parabolic trough. While using long, flat Fresnel mirrors, these collectors resemble parabolic troughs. Though less efficient, this technology is far less expensive to install. There are only a few plants as examples [20]. The heat exchanger tubes are often filled with synthetic oil or a molten salt solution to circulate through and absorb the heat generated by the mirrors. Typically heated to between 400 and 600°C [21] (Figure 3.10).

36 Thermal Energy Systems

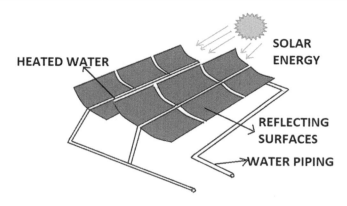

Figure 3.10 Parabolic collector [22].

3.4.2 Linear Fresnel Reflector

Giorgio Francia created this category of collectors. He created this technique in the 1960s in Genoa, Italy. Figure 3.11 shows a Fresnel reflector in action. This system's main benefit is that it uses less expensive flat curved reflectors instead of parabolic glass ones, which are more expensive. The structural requirements are also reduced by the near mounting to the ground of this type. The receiver tube of a linear Fresnel collector is surrounded by secondary reflectors, and it is based on several mirrors that focus light beams there. For this type of collector, there are various designs. A single absorber tube is all that is present in the basic design, however, newer models now support two or more tubes [23].

3.4.3 Parabolic Dish Reflector

The distributed-receiver system, sometimes referred to as a parabolic dish reflector, follows the sun in two directions during the day. This is necessary for

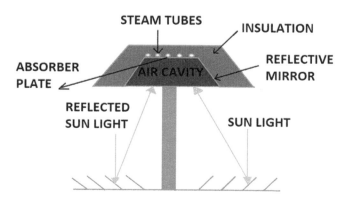

Figure 3.11 Linear Fresnel reflector [23].

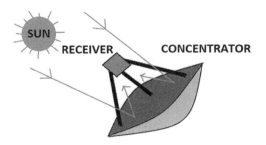

Figure 3.12 Parabolic dish reflector [24].

the dish architecture to reflect the beam to the thermal receiver. The radiant energy of the solar energy is captured by the receiver and converted into thermal energy in a flowing fluid. The thermal energy can either be converted into electricity by an engine generator attached to the receiver or it can be directly converted into electricity by pipes leading to a central power-conversion system. These collectors, which are the most effective of all collector systems, have a temperature limit of roughly 1500°C. They are very effective at absorbing thermal energy and converting it into power [24] (Figure 3.12).

3.4.4 Heliostat Field Collector

The primary focus in this context is on these types of solar power facilities. Heliostats are tracking mirrors that are located around towers and have a receiver on their heads. They are slightly concave. A schematic depiction of the heliostat field collector is presented in Figure 3.13 [25]. Large volumes of solar energy would be concentrated in this plant's receiving cavity and supplied there to power a steam generator, which would then produce steam at extremely high temperatures and pressures. Heat energy is taken up by the receiver and transmitted to a moving fluid where it can be stored and used to produce electricity at night. The range of this collector's concentration ratios is 300–1500 [26]. These kinds of plants occupy between 50 and 150 m^2 of space, and occasionally they make use of thermal storage systems. In some cases, hybrid plants are developed using the thermal

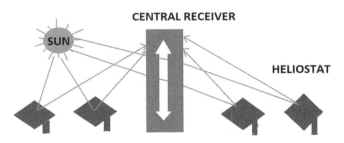

Figure 3.13 Heliostat field collector [25].

38 Thermal Energy Systems

storage system with both solar and fossil fuel energy. Between 200 and 1000 kW/m², the typical solar flux values entered into the receiver. It can operate at temperatures of up to 1500°C thanks to the high flux. The most common cycles employed in heliostat plants are Rankine and Brayton cycles, with the latter being used at higher temperatures [27].

The heliostats could be positioned either just north or just south of the reception tower depending on where the facility is located (in the northern hemisphere) (in the southern hemisphere). In addition, advanced systems may substitute air for water, steam, liquid sodium, or molten nitrate salt as the heat transfer fluid (sodium). The heat-collection medium fluid in more sophisticated systems should be oil mixed with broken rock, molten nitrate salt, liquid sodium, or ceramic bricks. Steam was used as the heat transfer medium in the Solar One early tower power plants, which presented a number of difficulties, including storage and continuous turbine operation [28]. Solar One was modified to Solar Two in order to address these issues, and this model used molten salt and even air as the medium fluid. In the Mojave Desert, an actual tower power collector known as Solar Two is depicted in Figure 3.13 [29–31].

3.5 DISCUSSION

Solar energy is freely available and eco-friendly. The globe receives a huge amount of this energy throughout the year. This energy if harnessed properly can help in sufficing the demands of the increasing population. This chapter has summarized all the solar collectors that can harness this energy from the sun and later store it for further use. The chapter has covered the solar panel varieties, how do they function, and numerous uses and strategies for promoting the advantages of the solar energy.

REFERENCES

1. Boes, E.C. *Fundamentals of solar radiation*. Albuquerque, NM (United States): Sandia National Lab. (SNL-NM), 1979 Dec 1.
2. Shafieian, A., Khiadani, M., Nosrati, A. A review of latest developments, progress, and applications of heat pipe solar collectors. *Renewable and Sustainable Energy Reviews* 2018 Nov 1; 95: 273–304.
3. Singh, R., Kumar, S., Hasan, M., Khan, M., Tiwari, G. Performance of a solar still integrated with evacuated tube collector in natural mode. *Desalination* 2013; 318: 25–33.
4. Meena, C.S., Kumar, A., Roy, S., Cannavale, A., Ghosh, A. Review on boiling heat transfer enhancement techniques. *Energies* 2022; 15: 5759. doi: 10.3390/en15155759
5. Meena, C.S., Kumar, A., Jain, S., Rehman, A.U., Mishra, S., Sharma, N.K., Bajaj, M., Shafiq, M., Eldin, E.T. Innovation in green building sector for sustainable future. *Energies* 2022; 15: 6631. doi: 10.3390/en15186631

6. Tiwari, G.N. Solar energy. *Fundamentals, design, modeling and applications*. New Delhi: Alpha Science International Ltd, 2006. p. 525.
7. Weiss, W., Themessl, A. *Training course – Solar water heating*. Latvia – Baltic States – Helsinki: Solpros AY, 1996. p. 55.
8. Direct Thermal Conversion and Storage [online] [viewed 2007 Nov 25]. Available: www.osti.gov/accomplishments/pdf/DE06877213/10.pdf
9. Jesko, Ž. Classification of solar collectors. *rN* 2008; 1(21): 21.
10. Yogi Goswami, D., Kreith, F., Kreider, J.F. *Principles of solar engineering*. New York: Taylor & Francis Group, 2000. p. 694.
11. Dabiri, S, Rahimi, M.F. Basic introduction of solar collectors and energy and exergy analysis of a heliostat plant. In The 3rd International Conference and Exhibition on Solar Energy, 2016, University of Tehran, Tehran, Iran
12. Duffie, J.A., Beckman, W.A. *Solar engineering of thermal processes*. New Jersey: John Wiley & Sons, Inc, 2006. p. 908.
13. Blanco, J., et al. Compound parabolic concentrator technology development to commercial solar detoxification applications. *Solar Energy* 1999; 67(4): 317–330.
14. Gajic, M., et al. Modeling reflection loss from an evacuated tube inside a compound parabolic concentrator with a cylindrical receiver. *Optics Express* 2015; 23(11): A493–A501.
15. Meena, C.S., Prajapati, A.N., Kumar, A., Kumar, M. Utilization of solar energy for water heating application to improve building energy efficiency: An experimental study. *Buildings* 2022; 12: 2166. 10.3390/buildings12122166.
16. Dawn, S., et al. Recent developments of solar energy in India: Perspectives, strategies and future goals. *Renewable and Sustainable Energy Reviews* 2016; 62: 215–235.
17. Goonan, T.G. The lifecycle of silver in the United States in 2009. US Department of the Interior. *US Geological Survey* 2013.
18. Singh, V.P., Jain, S., Karn, A., Dwivedi, G., Kumar, A., Mishra, S., Sharma, N.K., Bajaj, M., Zawbaa, H.M., Kamel, S. Heat transfer and friction factor correlations development for double pass solar air heater artificially roughened with perforated multi-V ribs. *Case Studies in Thermal Engineering* 2022; 39: 102461, ISSN 2214-157X. doi: 10.1016/j.csite.2022.102461
19. Singh, V.P., Jain, S., Karn, A., Kumar, A., Dwivedi, G., Meena, C.S., Cozzolino, R. Mathematical modeling of efficiency evaluation of double-pass parallel flow solar air heater. *Sustainability* 2022; 14: 10535. doi: 10.3390/su141710535
20. Boyle, G. *Renewable energy: Power for a sustainable future*. Milton Keynes: Oxford University Press, Oxford in association with The Open University, 1996.
21. Dobriyal, R., Negi, P., Sengar, N., Singh, D.B. A brief review on solar flat plate collector by incorporating the effect of nanofluid. *Materials Today: Proceedings* 2020 Jan 1; 21: 1653–1658.
22. Singh, V.P., Jain, S., Karn, A., Dwivedi, G., Alam, T., Kumar, A. Experimental assessment of variation in open area ratio on thermohydraulic performance of parallel flow solar air heater. *Arabian Journal for Science and Engineering* 2022. 10.1007/s13369-022-07525-7

23. Gouthamraj, K., Rani, K.J., Satyanarayana, G. Design and analysis of rooftop linear Fresnel reflector solar concentrator. *International Journal of Engineering and Innovation Technology* 2013; 2(11): 66–69.
24. Singh, V.P., Jain, S., Karn, A., Kumar, A., Dwivedi, G., Meena, C.S., Dutt, N., Ghosh, A. Recent developments and advancements in solar air heaters: A detailed review. *Sustainability* 2022; 14: 12149. doi: 10.3390/su141912149
25. Efficiency, E. Renewable energy technology characterizations. *Topical Report TR-109496, Energy Efficiency and Renewable Energy (EERE)*, US Department of Energy, 1997.
26. Sanchez, M., Romero, M. Methodology for generation of heliostat field layout in central receiver systems based on yearly normalized energy surfaces. *Solar Energy* 2006; 80(7): 861–874.
27. Besarati, S.M., Goswami, D.Y., Stefanakos, E.K. Optimal heliostat aiming strategy for uniform distribution of heat flux on the receiver of a solar power tower plant. *Energy Conversion and Management* 2014; 84: 234–243.
28. Singh, V.P., Jain, S., Kumar, A. Establishment of correlations for the thermo-hydraulic parameters due to perforation in a multi-V rib roughened single pass solar air heater. Experimental Heat Transfer 2022. doi: 10.1080/0891 6152.2022.2064940
29. Kushwaha, P.K., Sharma, N.K., Kumar, A., Meena, C.S. Recent advancements in augmentation of solar water heaters using nanocomposites with PCM: Past, present & future. *Buildings* 2023; 13:79. 10.3390/buildings13010079
30. Medrano, M., et al. State of the art on high-temperature thermal energy storage for power generation. Part 2—Case studies. *Renewable and Sustainable Energy Reviews* 2010; 14(1): 56–72.
31. Dutt, N., Binjola, A., Hedau, A.J., Kumar, A., Singh, V.P., Meena, C.S. Comparison of CFD results of smooth air duct with experimental and available equations in literature. *International Journal of Energy Resources Applications (IJERA)* 2022; 1(1): 40–47. doi: 10.56896/IJERA.2022.1.1.006

Chapter 4

Optimization of Solar Collector System Based on Different Nanofluids

Osama Khan, S. Mojahid Ul Islam, Tauseef Hassan, Hozaifa Ahmad, Md Adib Ur Rahman, and Ashok K. Dewangan

CONTENTS

4.1 Introduction 41
4.2 Material and Method.. 43
    4.2.1 Experimental Arrangement... 43
    4.2.2 Preparation of Nanofluids 44
    4.2.3 Evaluation of Uncertainty 45
4.3 Methodology... 45
    4.3.1 Variation of Operating Conditions 45
    4.3.2 Analytic Hierarchy Process (AHP) Analysis 46
    4.3.3 TOPSIS Assignment... 47
4.4 Result and Discussion 48
4.5 Conclusion 52
References.. 54

4.1 INTRODUCTION

The ever-changing climatic conditions have been an additional parameter to boost the solar industry, especially in tropical countries where relative humidity and ambient temperature are on the higher side. The population is entirely dependent on renewable industry and can be understood by the fact that they opt for places which are equipped with such cooling or heating systems. To counter extreme living conditions, solar energy potential is explored for multiple devices which require high power or energy to cool and heat the incoming stream of air. Researchers in past have been successful in designing systems which provide maximum efficiency at the lowest possible costs. Since several equipments are associated with changing conditions of ambient air, ample losses might be accompanied while designing the equipment. In general, if the equipment deviates even slightly from its original design, then it might affect the performance of the whole system, making it ineffective and inefficient. Henceforth, the primary aim of the researchers must be directed towards furnishing the best possible

DOI: 10.1201/9781003395768-4

conditions while reducing the operating and maintenance cost associated with the system.

In the past, solar collectors have proved to be a potent source of energy production probably due to their cleaner nature, easy availability, lower exhaust emissions, easier installation and no fuel requirement. Although solar collector-based power generation looks like a cleaner, feasible and viable option, there are certain complications which need to be addressed to furnish sustainable results. One such difficulty encountered in solar collectors is the generation of excessive heat during the prolonged radiation absorption period. This deviates the operating conditions creating unwanted complications in the system, eventually resulting in lowered efficiency of the power conversion process. Especially in flat plate collectors, the surface temperature of the panel increases by several folds, thereby harming its capability of conversion and acceptance of solar radiation. In order to avoid heat up of solar panels multiple fluids are recirculated on the collectors to lower its temperature in the acquired range. Nanofluids due to increased surface and higher thermal conductivity carry a higher amount of heat energy thereby transferring to another domain requiring heat for a co-generation system, henceforth simultaneously producing power and heat together. In this study, multiple nanofluids are selected for an analysis whose operation and performance capabilities are compared with one another [1–11].

Solar collectors (SC) can be made more effective by spreading layers of various coatings on the absorber system of the solar equipment, modifying the mass flow rate of the applied fluid [12–25], and varying the original design by angling the collector towards different angles towards the sun. Nevertheless, this system demands advancements, such as the addition of a novel liquid type as a typical operating fluid, designated as "nanofluid," to boost the thermal efficiency from the base plate to the operating fluid. A uniform nanofluid is essentially applied as a working fluid in a thermal system involving transfer of heat. As nano-scale atoms of nano-additive spread in the base fluid, physio-chemical properties such as buoyancy, gravity and other body forces have no major effect on the colloidal consistency of the solution. Dynamic viscosity therefore governs the majority of base fluid uniformity. Covalent or quasi (surfactant) nanoparticle is mixed to improve the long-term consistency of the nanoparticles in the original fluid. The use of electrostatically and quasi-synthesized nanomaterial in photovoltaic solar collectors or receivers is seldom explored as far as the wide literature of solar energy is concerned. In this research, the proposed study begins with the creation of nanomaterials by applying molecular polymerization technique with nanostructure as well as surfactant-treated technique with multiple nanocrystals. Using metallic compounds, electrolytic calcium hexametaphosphate has been applied like organic surfactants. The physio-chemical properties of the developed nanofluids have been investigated as well as correlated and compared to theoretical models. The evaluated values were found to be in accordance with the global ASHRAE

Standards. The uncertainties of the measured equipment were also taken into consideration in order to obtain foolproof values with a minimum error rate.

Researchers have studied the statistical method to prognosticate solar radiation through climate-forecasting representations [26]. These models were explored on the basis of space and temperature. Researcher has proposed a method for the estimation of global solar radiation on the tilted surface in 5-min steps but still remains a difficult step to measure due to high cost and maintenance requirements [27]. Yadav et al. [28] presented frameworks for small periods and longer periods of solar radiation which worked on several climatic variables and were compared on accuracy and cost-effectiveness. Ahmad and Tiwari [29–40] estimated optimum tilt angles for India by employing Liu & Jordan Model capable of estimating the yearly tilt angle which is calculated approximately equal to the latitude.

From previous studies established above, it is quite evident that the impact of higher temperature deteriorates the efficiency of the solar panel system. On the other hand, the presence of nanofluids as a recirculating medium instead of simple water effectively cools the whole system. thereby enhancing its overall performance parameters. It is quite essential to identify the best nanofluid in multiple sets available in the markets. so far single nanofluids have been used in research not drawing a comparison among different nanofluids based on performance parameters. The decision to identify and select the best nanofluid for a flat plate collector among the rest is researched in this chapter for the first time using multi-criteria decision-making (MCDM) processes, thereby marking its novelty.

After analysing a vast literature survey exploring MCDM in solar applications, the authors can interpret that seldom studies were found in the area of selection of nanofluids in flat plate collectors using MCDM techniques. This work will provide future researchers with a framework to analyse different nanofluids by using the hybrid AHP-TOPSIS model. The contemporary study selects various operating parameters and then assigns weights to them, thereby prioritizing specific parameters which have a bigger role in performance parameters. The research considers seven parameters namely pressure loss, viscosity, density, specific heat, volumetric flow rate, power and mass flow rate. The different nanofluids are ranked with the aid of TOPSIS method. the nanofluids considered for the study are graphene oxide, deionized water, silicon oxide and zinc oxide.

4.2 MATERIAL AND METHOD

4.2.1 Experimental Arrangement

The whole setup of recirculating nanofluids was furnished in the present setup as shown in Figure 4.1. The complete set of equipment used in

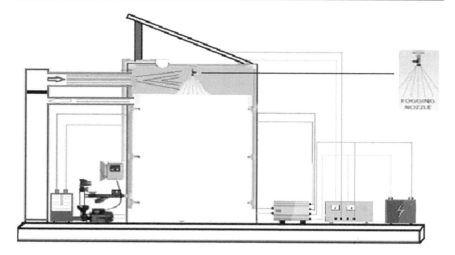

Figure 4.1 Experimental setup.

designing the solar setup is provided in Figure 4.1. The flat plate collectors are aligned in series capable of producing 500 W of power. The tubes used for recirculating liquid are made up of aluminium nitride material and are placed adjacent to the flat plate collectors so as to provide a uniform cooling process. A small pump is used to process and flow nanofluid around the system with 0.4 kg/s stable mass flow rate.

Temperature coefficients of nanofluids are recorded at regular intervals during entry and exit points of flat plate collectors. The PV panel comprises a generator solar array in addition to junctions and protector parts. The solar cells attached at the top further are an amalgamation product of p-n junctions attached to silicon-based semiconductor sheet material. The setup further comprises an inverter of 700 VA which transforms DC to AC, as the pump requirements are specified as AC.

4.2.2 Preparation of Nanofluids

The reactive hydroxide compounds were introduced mostly over layers of NPs by a mild alkaline solution. The primal initial procedure is to create a 1:3 mixture containing nitric acid (HNO_3) and sulphuric acid (H_2SO_4). Afterwards, 5 g NPs (750 m^2/g) was incorporated into another high alkaline fluid. This simultaneous resulting technique has been used for soaking 3 hours at the same compact form frequency. This same suspended mixture has been poured into a thermoplastic glass beaker. This sample solution gets dried at the ambient level and then passed over using a polytetrafluoroethylene cell wall before being rinsed multiple times using distilling liquid (DI water) to achieve balanced pH values. Afterwards, this same experiment was then left to dry inside a convection chamber. Silicon dioxide (SiO_2), a nano-additive with a purity of 99.5% having crystal dimensions varying from 10 to 20 nm, was used in the analysis. later on, distilled aqua was intermixed with nano-additive to produce nanofluid.

4.2.3 Evaluation of Uncertainty

Often the data acquired by experimentation comprises of internal defects and inconsistencies. For any research work to be accurate and precise, it is essential we remove these discrepancies in the system. Henceforth, the current research model estimates and records all the uncertainties in the system by using Holman's model of uncertainty. According to above study would correspondingly describe the performance of the system by incorporating the Holman principle of uncertainty which estimates the overall difference in values demonstrated by the expected readings and actual readings.

The total percentage of uncertainty is determined in this experiment by applying Holman's equation provided:

The total percentage uncertainty = square root of [(uncertainty in T − type thermocouples)^2

+ (uncertainty in Flow meter)^2

+ (uncertainty in Pressure transducer)^2

+ (uncertainty in Voltage measurement)^2

+ (uncertainty in Current measurement)^2

+ (uncertainty in SiS sensor)^2

+(uncertainty in Power Temp Coefficient)^2]

The total percentage uncertainty = Square root of[$(0.4)^2 + (0.003)^2 + (1.3)^2 + (0.03)^2$

$+ (0.12)^2 + (2)^2 + (0.23)^2$]

The total percentage uncertainty = ±3.75%

As a result, the measured FPSC efficiency value remained in the zone of 3.75% which is an acceptable range. In this research, synthesized nanostructures and thin film nanoparticles are used to evaluate the thermodynamic efficiency of a flat plate concentrator which previously has never been explored as far as authors' knowledge comprehends, thereby establishing its novelty.

4.3 METHODOLOGY

4.3.1 Variation of Operating Conditions

Four variants of nanofluids are selected on different operating conditions in the same concentration for a flat plate collector. The concentrations for GO, ZnO and SiO_2 nanofluids were the same for each experiment so as to achieve a stabilized cooling process. The operating parameters chosen for

46 Thermal Energy Systems

the study are pressure loss, viscosity, density, specific heat, volumetric flow rate, power and mass flow rate. The nanofluids to be ranked are graphene oxide (GO), zinc oxide (ZO), silicon oxide (SO) and deionized water.

4.3.2 Analytic Hierarchy Process (AHP) Analysis

AHP technique is capable of contributing to problems where datasets are complex, unrelated, and require multi-performance criteria. This permits the decision-makers in evaluating significant challenges and segregates them into feasible and cost-effective sub-systems.

The following are the quick overview of AHP steps: -

Step 1. Initially, identify the goals to accomplish, subsequently restricting possible choices.

Identifying parameters and judging them based on a quantifiable characteristic requires the practical expertise of a seasoned expert who can present a proper explanation and tabulation of the available options.

Step 2. A thorough assessment needs to be performed, which is classified into two sections: (i) Among attributes of the problem (ii) Between many alternative solutions that fulfil every prerequisite.

Experts in the required field have developed matrices which analyse attributes on a primary scale ranging from 1 to 5.

The number of interrelated vectors is calculated using the formula (nxn), where n specifies the number of attributes.

Step 3. Let Xij indicates the ith factor's priority ranking in comparison to the jth factor. After that, $Xji = 1/X_{ij}$.

Step 4. Preparation of a standardized pair-wise matrix is created for the applied system in the following sub-steps:
   a. Compute the total of all columns.
   b. Deduct the sum of the derived column from each element of the matrices.
   c. For getting the relative weights, calculate the number of rows.

Step 5. By applying equation (4.11), we get the Evaluation matrix, highest Evaluation result, and Criterion index (*CI*)

$$CI = \frac{\lambda \max - n}{n - 1} \qquad (4.1)$$

The Eigenvector of the conjoint analysis matrices is maximum, and the number of criteria is n.

Step 6. By using equation (4.1), the consistency ratio (CR) is calculated (4.2), where RI means random index,

$$CR = \frac{CI}{RI} \tag{4.2}$$

4.3.3 TOPSIS Assignment

The multiple attribute performance models furnished by TOPSIS method was adopted in this research to enhance the performance of the solar prediction models, which includes the following steps:

Step 1. Gather data from stakeholders regarding the essential evaluations of the outcome responses in linguistic terms such as extremely low, low, average, high, and extremely high from the experts.

Step 2. Transforming various linguistic values into numeral valuations.

$X_{abN} = (l_{abN})$ where, $a = 1, 2, 3\ldots\ldots\ldots m$; $b = 1, 2, 3\ldots\ldots\ldots n$, where,

$$a = \min\{l_{abN}\}, \quad b = \frac{1}{N} \sum_{N=1}^{N} P_{abN}, \quad c = \max(u_{abN}) \tag{4.3}$$

Step 3. Evaluate the outcome responses for the combined weights.

$$B = [P_{ij}]_{mxn} \tag{4.4}$$

Here, $i = 1, 2, 3, \ldots ., m$; $j = 1, 2, 3, \ldots ., n$

$$P_{ij} = \left(\frac{a_{ij}}{c_j^*}\right); \quad c_j^* = \max c_{ij} \tag{4.5}$$

$$P_{ij} = \left(\frac{a_j^-}{c_{ij}}\right); \quad a_j^* = \min a_{ij} \tag{4.6}$$

Step 4. Standardize the overall output matrices.

$$V = [v_{ij}]_{mxn} \text{ where } i = 1, 2, 3, \ldots, m; \quad j = 1, 2, 3, \ldots, n \tag{4.7}$$

48 Thermal Energy Systems

Here, $v_{ij} = p_{ij}(\times)w_j$ (4.8)

Step 5. Compute the standardized weighted matrices.

$A^+ = \{v_1^+, \ldots\ldots, v_n^+\}$ where (4.9)

$v_j^+ = \left\{\max(v_{ij})\,IFj \in J; \quad \min v_{ij}\,IFj \in J'\right\}, j = 1, \ 2, \ 3, \ \ldots, \ n$ (4.10)

$A^- = \{v_1^-, \ldots\ldots, v_n^-\}$ where

$v_j^- = \left\{\max(v_{ij})\,IFj \in J; \quad \min v_{ij}\,IFj \in J'\right\}, j = 1, \ 2, \ 3, \ \ldots, \ n$ (4.11)

Step 6. Establish the optimal solutions that are both positive and negative.

$d_i^+ = \left\{\sum_{j=1}^{n}\left(v_{ij} - v_{ij}^+\right)\right\}^{1/2}; \quad i = 1, \ 2, \ \ldots, m$ (4.12)

$d_i^- = \left\{\sum_{j=1}^{n}\left(v_{ij} - v_{ij}^-\right)\right\}^{1/2}; \quad i = 1, \ 2, \ \ldots, m$ (4.13)

Step 7. Compute the differences between the actual data collected from the ideal both positive and negative.

$CC_i = \dfrac{d_i^-}{d_i^- + d_i^+}; \quad i = 1, \ 2, \ \ldots, n$ (4.14)

Step 8. Estimate the closeness coefficient (CC) data and evaluate the pre-experimental studies based on them, starting with the research having the maximum CC value marked with the highest rank. The rank degrades with decreasing CC value. Table 4.1 illustrates the operating parameters assigned to the significance levels.

4.4 RESULT AND DISCUSSION

This section in particular explores the operating conditions of various nanofluids in actual values, further providing weights to all constraints with the aid of AHP model. Later on, TOPSIS analysis introduces the ranking analysis based on the above-derived weights, thereby establishing the best working and worst working nanofluid for a flat plate collector. Evidently,

Table 4.1 Actual values assigned to the operating conditions of the significance levels

Type of fluid parameter	Pressure loss (kPa)	Viscosity (m Pa s)	Density (kg/m^3)	Specific heat (J/kg K)	Volumetric flow rate (m^3/s)	Power (W)	Mass flow rate (kg/s)
Graphene oxide	380.6	1.21	1141.8	3432	2.926	12.65	0.0308
Deionized water	473	2.497	1502.6	2572.9	2.079	11.176	0.02838
Silicon oxide	374	1.551	1311.2	2989.8	1.837	7.81	0.02189
Zinc oxide	354.2	1.287	1227.6	3103.1	2.277	9.163	0.0253

50 Thermal Energy Systems

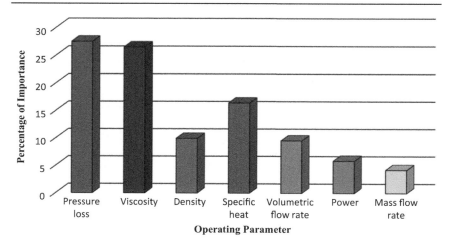

Figure 4.2 Priority of operating parameters.

from the analysis pressure loss came out to be the most essential operating parameter whereas mass flow rate came to be least impactful in case of a flat plate collector. Based on the analysis, the sequence of operating parameters based on priority is shown in Figure 4.2. The sequence of importance is from highest to lowest: pressure loss, viscosity, specific heat, volumetric flow rate, density, power and mass flow rate. During the TOPSIS analysis, each nanofluid is ranked based on the priority of weights and their value assigned. Each nanofluid is ranked on the basis of the closeness coefficient which is a measure of the score attained by each nanofluid.

The closeness coefficient as calculated in 5 steps and is depicted in Figure 4.3. Regardless of the operating characteristic of the yield responses,

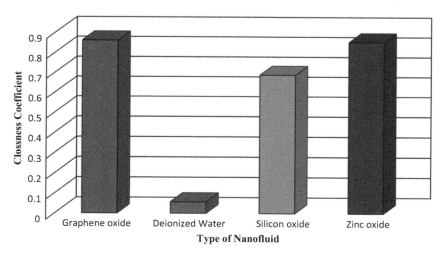

Figure 4.3 Closeness coefficient of various nanofluids.

a general trend that is followed in prior research is that a higher value is favourable whereas a lower value is not beneficial. Among these characteristics, the most impactful parameters are justified by this hybrid model. As per the closeness coefficient, graphene oxide nanofluid registers the highest value, thereby attaining the top rank while the lowest value was attained in the case of deionized water registering the lowest rank.

The steady dissolvable nanofluid used as an operating liquid was created through molecular deposition of metal nanosheets and semi-copolymerization (emulsifier) of NiO-based transition metals. An important factor in the convective heat transmission is the thermal conductivity of any nano-additive mixture. KD2-pro is employed to determine the thermal conductivity of hybrid nano-additive liquid and base fluids, gaining enhancements in NPs, silicon dioxide, and zinc, over deionized aqua. Graphene, Zinc, and CuO working liquid showed improvements in density, specific heat, and permeability. By improving the heat transfer ratio of solar collectors, which enhanced its overall efficiencies. When mixed with DI water, the thermal efficiency of the absorber surface has been evaluated by all liquid nanoparticles. In conclusion, all nanoparticle solutions performed better than deionized liquid for flat plate collectors' thermal efficiency, and they can be employed as an additional operating liquid. Figure 4.4 shows the impact percentage of various operating parameters in a flat plate collector system.

The readings of NP and water under different volume fluid conditions and density contents were recorded, thereby demonstrating that the overall thermal dissipation ratio of nanoparticles is larger than that of liquid and rises with fluid velocity. As earlier stated, deionised water has inferior

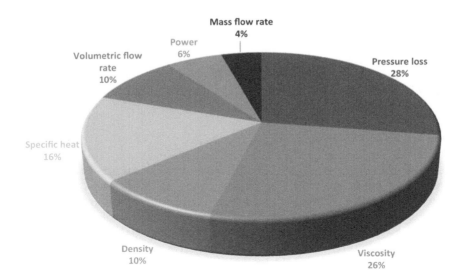

Figure 4.4 Percentage distribution of operating parameters.

52 Thermal Energy Systems

thermal efficiency as compared to nanofluids. The convection heat transmission ratio for graphene NP was found to be higher as compared to simple water. Thus, the heating capacity of a photovoltaic collector could be increased by employing GNP/water nanofluids. Due to the GNP/water nanofluid's improved thermal conductance and the rising tube's interior wall thickness eventually decreased heat resistivity. The rising tube's small heat transfer surface on the walls is principally responsible for the abrupt increase in the heat transfer coefficient. Also, carbon-based composites like graphene and CNT lead to smaller heat boundary layers. Convective thermal transfer coefficients generally rise with subsequent increments in specific surface areas as well as the Brownian motion of nanomaterials. Furthermore, the convective heat transfer coefficient can be improved for a simultaneous increment in Re evidently seen for nanofluids as well as DI water. As a result, adding nanoparticles to the base fluid boosts the thermal performance of FPSC. The improvement in the thermal effectiveness of the solar collector is due to the fact that the specific heat capacity (Cp) of nanofluids is lower than the base fluid (water), but the temperature differential (DT) for nanofluids is noticeably larger than the base fluid. The enhanced heat conductance of nanofluids is the primary basis of this increment in temperature difference which can be validated by past studies. The computed results of TW and AP can be observed to decrease as flow rate rises, and corresponding values for water-based ZnO and SiO_2 nanofluids are lower than their equivalent water values. Due to the advancements in the original fluid by addition of ZnO and SiO_2 nanoparticles' thermal conductivities, such a drop will lead to an increase in the material's heat transfer coefficient. As justified in previous studies, the application of nanofluids frequently exhibits better convective heat transfer coefficients in laminar and turbulent environments as well as greater thermal conductivities than their corresponding base fluids. The exergy effectiveness of flat plate collectors for nanomaterials like functionalized carbon and metal oxide-based nanofluids at various flow rates. Additionally, all of the nanofluids tested in the current study had higher exergy efficiency when compared to DI water.

4.5 CONCLUSION

In present times, the emergence and acceptance of various non-conventional energy technologies are probably due to the awareness developed in the new generation regarding the rising global temperature due to the increase in greenhouse gas emissions emitted during the employment of conventional power production systems. Reports suggest that at this pace the overall temperature of the globe might rise 1.5 times between the years 2030 and 2050 which would have devastating effects on the living biosphere creating an imbalance in ecosystem. In this research, the heating problem of flat plate collectors is discussed along with design operating parameters. The cooling is achieved by recirculating multiple nanofluids through the experimental setup measured at various intervals. The operating parameters were weighed and prioritized based on AHP technique and TOPSIS method was used to rank the nanofluids among them.

It was detected that the zenith value of closeness coefficient was accomplished for graphene oxide, consequently, it comes out to be the significant input parameter nanofluid that affects the multi-performance properties of flat plate collectors. This research established that the AHP-TOPSIS method was quite effective and operative in solving problems related to flat plate collector multi-optimization.

As contrasted to DI water, the experimental thermal performance of flat plate collectors versus lowered temperature factors at a weight content equal and various mass flow rate. The water-based NP nanofluids exhibit superior performance as compared to nanofluids created in this study, in comparison with that of water. It exhibits the best thermal effectiveness of FPSC, equivalent. As a consequence of a percent increase in FP thermal performance attained at a stream rate, dependent nanofluids can be ordered as GO > ZnO > SiO_2 > DW. A relatively similar pattern was seen for the higher mass flow rate, although the associated thermal efficiency percentage increases.

Any research has some inaccuracy in some manner. The following list outlines the disadvantages of the current analysis, which is similar to all previous analyses.

- Fewer nanocrystalline materials were employed to create mixtures of biomass fuels.
- During the simulations, hardly four control factors have been taken into account, which would be insufficient for other large systems.
- A small range of effectiveness parameters was taken into account.
- The method employed for various functional attribute evaluations as in current research primarily allows for the most appropriate initial value selection within the evaluated range.

In order to establish important advancements in the field of science, researchers should organize and carry out studies to examine previously undiscovered events within the light of original study limits.

The following ideas could be incorporated into the ongoing study:

- The testing of several other nanomaterials and quasi-oriented reinforced composites could be used to produce a more thorough comparative experimental study.
- Various control variables having multiple points could be taken into account.
- Maximum effectiveness attributes might also be evaluated, which would include additional operating factors like mixing rate and mass flow rate.

- In scientific research, the entire comprehensive approach of studies might also be utilized.
- In the coming years, investigations could lead to innovative evolutionary algorithm methodologies

REFERENCES

1. Ahmad, S., Alam, M.T., Bilal, M., Khan, O., & Khan, M.Z. (2022). Analytical Modelling of HVAC-IoT Systems with the Aid of UVGI and Solar Energy Harvesting. In: Agarwal, D., Verma, K., Urooj, S. (eds). *Energy Harvesting Enabling IoT Transformations*, 65–80.Boca Raton, FL, USA: CRC Press.
2. Parvez, M., Khan, M.E., Khalid, F., Khan, O., & Akram, W. (2021). A Novel Energy and Exergy Assessments of Solar Operated Combined Power and Absorption Refrigeration Cogeneration Cycle. In: Patel, N., Bhoi, A.K., Padmanaban, S., and Holm-Nielsen, J.B. (eds). *Electric Vehicles. Green Energy and Technology.*Singapore: Springer. 10.1007/978-981-15-9251-5_13
3. Shukla, N.K., Rangnekar, S., & Sudhakar, S. (2015). Comparative study of isotropic and anisotropic sky models to estimate solar radiation incident on tilted surface: A case study for Bhopal India. *Energy Reports*, 1, 96–103.
4. Meena, C.S., Kumar, A., Roy, S., Cannavale, A., & Ghosh, A. (2022). Review on boiling heat transfer enhancement techniques. *Energies*, 15, 5759. doi: 10.3390/en15155759
5. Mani, A. (1980). *Handbook of Solar Radiation Data for India*, New Delhi: Allied Publishers Pvt. Ltd.
6. Meraj, M., Khan, M.E., Tiwari, G.N., & Khan, O. (2019). Optimization of Electrical Power of Solar Cell of Photovoltaic Module for a Given Peak Power and Photovoltaic Module Area. In: Saha, P., Subbarao, P., and Sikarwar, B. (eds). *Advances in Fluid and Thermal Engineering. Lecture Notes in Mechanical Engineering*. Singapore: Springer. 10.1007/978-981-13-6416-7_40
7. Khan, S., Tomar, S., Fatima, M., & Khan, M.Z. (2022). Impact of artificial intelligent and industry 4.0 based products on consumer behaviour characteristics: A meta-analysis-based review. *Sustainable Operations and Computers*, 3, 218–225.
8. Fatima, M., Sherwani, N.U.K., Khan, S., & Khan, M.Z. (2022). Assessing and predicting operation variables for doctors employing industry 4.0 in health care industry using an adaptive neuro-fuzzy inference system (ANFIS) approach. *Sustainable Operations and Computers*, 3, 286–295.
9. Singh, V.P., Jain, S., Karn, A., Kumar, A., Dwivedi, G., Meena, C.S., Dutt, N., & Ghosh, A. (2022). Recent developments and advancements in solar air heaters: A detailed review. *Sustainability*, 14, 12149. doi: 10.3390/su141 912149
10. Khan, O., Khan, M.Z., Khan, M.E., Goyal, A., Bhatt, B.K., Khan, A., & Parvez, M. (2021). Experimental analysis of solar powered disinfection tunnel mist spray system for coronavirus prevention in public and remote places. *Materials Today: Proceedings*, 46(15), 6852–6858.

11. Parvez, M., Khalid, F., & Khan, O. (2020). Thermodynamic performance assessment of solar-based combined power and absorption refrigeration cycle. *International Journal of Exergy*, 31(3), 232–248.
12. Seraj, M.M., Khan, O., Khan, M.Z., Parvez, M., Bhatt, B.K., Ullah, A., & Alam, M.T. (2022). Analytical research of artificial intelligent models for machining industry under varying environmental strategies: An industry 4.0 approach. *Sustainable Operations and Computers*, 3, 176–187.
13. Parvez, M., & Khan, O. (2020). Parametric simulation of biomass integrated gasification combined cycle (BIGCC) power plant using three different biomass materials. *Biomass Conversion and Biorefinery*, 10(4), 803–812.
14. Yadav, A.K., Khan, O., & Khan, M.E. (2018). Utilization of high FFA landfill waste (leachates) as a feedstock for sustainable biodiesel production: Its characterization and engine performance evaluation. *Environmental Science and Pollution Research*, 25(32), 32312–32320.
15. Khan, O., Khan, M.Z., Khan, E., Bhatt, B.K., Afzal, A., Ağbulut, U., & Shaik S. (2022). An enhancement in diesel engine performance, combustion, and emission attributes fueled with *Eichhornia crassipes* oil and copper oxide nanoparticles at different injection pressures. *Energy Sources, Part A: Recovery, Utilization, and Environmental Effects*, 44(3), 6501–6522.
16. Khahro, F.S., Tabassum, K., Talpur, S., Alvi, B.M., Liao, X., & Dong, L. (2015). Evaluation of solar energy resources by establishing empirical models for diffuse solar radiation on tilted surface and analysis for optimum tilt angle for a prospective location in southern region of Sindh, Pakistan. *Electrical Power and Energy Systems*, 64, 1073–1080.
17. Khan, O., Khan, M.E., Parvez, M., Ahmed, K.A.A.R., & Ahmad, I. (2022). Extraction and Experimentation of Biodiesel Produced from Leachate Oils of Landfills Coupled with Nano-additives Aluminium Oxide and Copper Oxide on Diesel Engine. In: Khan, Z. H. (ed). *Nanomaterials for Innovative Energy Systems and Devices. Materials Horizons: From Nature to Nanomaterials*. Singapore: Springer. 10.1007/978-981-19-0553-7_8.
18. El-Sebaii, A., & Trabea, A. (2005). Estimation of global solar radiation on horizontal surfaces over Egypt. *Egypt Journal of Solids*, 28, 163–175.
19. Makade, G.R., Chakrabarti, S., & Jamil, B. (2019). Prediction of global solar radiation using a single empirical model for diversified locations across India. *Urban Climate*, 29, 100492.
20. Jacovides, C., Tymvios, F., Assimakopoulos, V., & Kaltsounides, N. (2006). Comparative study of various correlations in estimating hourly diffuse fraction of global solar radiation. *Renewable Energy*, 31, 2492–2504.
21. Mohammadi, B., & Moazenzadeh, R. (2021). Performance analysis of daily global solar radiation models in Peru by regression analysis. *Atmosphere*, 12, 389.
22. Yanni, B., Danyi, S., Zhenming, S., Yan, Z., & Ming, P. (2021). Application of Arma model in deformation monitoring and forecasting of anchorage foundation Pit[J]. *Journal of Engineering Geology*, 29(5), 1621–1631.
23. Singh, V.P., Jain, S., Karn, A., Dwivedi, G., Kumar, A., Mishra, S., Sharma, N.K., Bajaj, M., Zawbaa, H.M., & Kamel, S. (2022). Heat transfer and friction factor correlations development for double pass solar air heater artificially roughened with perforated multi-V ribs. *Case Studies in Thermal Engineering*, 39, 102461, ISSN 2214-157X, 10.1016/j.csite.2022.102461

24. Dey, S., Reang, N.M., Das, P.K., & Deb, M. (2021). Comparative study using RSM and ANN modelling for performance-emission prediction of CI engine fuelled with bio-diesohol blends: A fuzzy optimization approach. *Fuel*, 292, 120356. 10.1016/j.fuel.2021.120356

25. Ahmed, M., Masood, S., Ahmad, M., & El-Latif, A.A.A. (2021). Intelligent driver drowsiness detection for traffic safety based on multi CNN deep model and facial subsampling. *IEEE Transactions on Intelligent Transportation Systems*, 23(10), 19743–19752. 10.1109/TITS.2021.3134222

26. Scabbia, G., Sanfilippo, A., Perez-Astudillo, D., Bachour, D., & Fountoukis, C. (2022). Exploring the limits of machine learning in the prediction of solar radiation. In: Heggy, E., Bermudez, V., Vermeersch, M. (eds), *Sustainable Energy-Water-Environment Nexus in Deserts. Advances in Science, Technology & Innovation*. Cham: Springer.

27. Diez, F.J., Martínez-Rodríguez, A., Navas-Gracia, L.M., Chico-Santamarta, L., Correa-Guimaraes, A., & Andara, R. (2021). Estimation of the hourly global solar irradiation on the tilted and oriented plane of photovoltaic solar panels applied to greenhouse production. *Agronomy*, 11, 495.

28. Yadav, A.K., Malik, H., Chandel, S.S., Khan, I.A., Otaibi, S.A., & Alkhammash, H.K. (2021). Novel approach to investigate the influence of optimum tilt angle on minimum cost of energy-based maximum power generation and sizing of PV Systems: A case study of diverse climatic zones in India. *IEEE Access*, 9, 110103–110115.

29. Sharma, N., Tiwari, P.K., Ahmad, G., & Sharma, H. (2021). Optimum tilt and orientation angle determination with application of solar data. *2021 International Conference on Artificial Intelligence and Smart Systems (ICAIS)*, IEEE, 477–481.

30. Puah, B.K., Chong, L.W., Wong, Y.W., Begam, K.M., Khan, N., Juman, M.A., & Rajkumar, R.K. (2021). A regression unsupervised incremental learning algorithm for solar irradiance prediction. *Renewable Energy*, 164, 908–925.

31. Mohanty, S. (2014). ANFIS based prediction of monthly average global solar radiation over Bhubaneswar (State of Odisha). *International Journal of Ethics in Engineering & Management Education*, 5(1).

32. Voyant, C., Notton, G., Kalogirou, S., Nivet, M.L., Paoli, C., Motte, F., & Fouilloy, A. (2017). Machine learning methods for solar radiation forecasting: A review. *Renewable Energy*, 105, 569–582.

33. Bilal, M., Zeeshan, M., Alam, M.T., Bhatt, B.K., & Khan, O. (2022). Analysis and validation of an intelligent energy monitoring framework for a yacht powered system based on adaptive neuro-fuzzy inference system (ANFIS). *International Journal of Robotics Research and Development (IJRRD)*, 12 (1), 55–64.

34. Khan, O., Khan, M.E., Yadav, A.K., & Sharma, D. (2017). The ultrasonic-assisted optimization of biodiesel production from eucalyptus oil. *Energy Sources, Part A: Recovery, Utilization, and Environmental Effects*, 39(13), 1323–1331.

35. Khan, O., Khan, M.Z., Bhatt, B.K., Alam, M.T., & Tripathi, M. (2022). Multi-objective optimization of diesel engine performance, vibration and emission parameters employing blends of biodiesel, hydrogen and cerium oxide nanoparticles with the aid of response surface methodology approach.

International Journal of Hydrogen Energy, ISSN 0360-3199. 10.1016/j.ijhydene.2022.04.044

36. Song, Z., Ren, Z., Deng, Q., Kang, X., Zhou, M., Liu, D., & Chen, X. (2020). General model for estimating daily and monthly mean daily diffuse solar radiation in China's subtropical monsoon climatic zone. *Renewable Energy*, 145, 318–332.

37. Behrang, M.A., Assareh, E., Ghanbarzadeh, A., & Noghrehabadi, A.R. (2010). The potential of different artificial neural network (ANN) techniques in daily global solar radiation modeling based on meteorological data. *Solar Energy*, 84(8), 1468–1480.

38. Antonopoulos, V.Z., Papamichail, D.M., Aschonitis, V.G., & Antonopoulos, A.V. (2019). Solar radiation estimation methods using ANN and empirical models. *Computers and Electronics in Agriculture*, 160, 160–167.

39. Khan, O., Yadav, A.K., Khan, M.E., & Parvez, M. (2019). Characterization of bioethanol obtained from *Eichhornia Crassipes* plant; its emission and performance analysis on CI engine. *Energy Sources, Part A: Recovery, Utilization, and Environmental Effects*, 43, 1–11.

40. Khan, O., Khan, M.Z., Ahmad, N., Qamer, A., Alam, M.T., & Siddiqui, A.H. (2019). Performance and emission analysis on palm oil derived biodiesel coupled with Aluminium oxide nanoparticles. *Materials Today: Proceedings*, 46, 6781- 6786. ISSN 2214-7853, 10.1016/j.matpr.2021.04.338

Chapter 5

Advancements in Solar Thermal and Photovoltaic System

Bipasa Patra and Pragya Nema

CONTENTS

5.1 Overview 59
5.2 What Is Solar Thermal? 62
5.3 Principle of Working of Solar Thermal System 62
5.4 What Is Solar PV? 63
5.5 Recent Advances in Solar PV and Thermal Technologies 64
    5.5.1 Recent Advancements in the Solar Cell Materials 65
        5.5.1.1 Crystalline Configuration 66
        5.5.1.2 Thin Film Technology 66
        5.5.1.3 Solar PV Concentrated Technology 67
        5.5.1.4 Polymer and Organic Configuration 68
        5.5.1.5 Hybrid Configuration 68
        5.5.1.6 Dye-Sensitized Type 69
5.6 Advanced Applications and Trends in Solar PV 69
5.7 Conclusion 70
References 70

5.1 OVERVIEW

The solar PV system finds its core application in the power sector for the generation of electricity by employing solar photovoltaics (PVs). The arrangement includes the involvement of PV panels for absorption of the irradiation, and later conversion into energy. For industrial and residential applications, a solar inverter is employed for the conversion of DC to AC. The improvised tracking techniques and the proper application of MPPT or perturb and observed methods as per suitability adjust the panels for the maximized harnessing of the solar irradiations. The balance of the system and the solar array are two different aspects. The two types of configuration of solar power systems are solar thermal and PV systems. The major principle is the same for both types of systems namely solar PV and solar thermal. The motive is to absorb the raw form of energy from the sun

DOI: 10.1201/9781003395768-5

59

and reform and recreate it to an effectively usable form. The major differentiations are remarked as for the case of PV system by the generation of electric signals whereas the other module being thermal uses, the direct heat from the sun is used to heat water or air. Both systems hold demand in household applications. But if we refer to commercialization, the thought process favours return-investment calculations primarily and then the concerns related to sustainability. The systems may be used individually or jointly. The primitive technology involves the usage of thermal energy from the sun to heat water at a wider scenario. It must be noted that CSPs-concentrated solar power finds its importance in generation and as a promising edge in gaining research interests regarding storage patterns and advances in this domain.

The researchers have discovered that the renewable sources of energy are facing some issues to be implemented with lowered efficiency, greater installation and maintenance costs and geographical constraints regarding availability over and across the globe. Since the cost of such systems for the time being is high as compared to the existing fossil fuels and also the working with efficiency is a serious challenge, they are of lesser interest to the common man. This chapter focuses on the recent advances in solar thermal and PV systems that are able to deal with and overcome these issues to a quite greater extent. But for this, a detailed study of each type is needed first to explore the areas of expansion and further development in solar thermal and solar PV systems (Figure 5.1).

The thermal type withstands and sustains energy storage for longer hours as compared to other types. The CSP is utilized for the areas with greater energy demands. The emerging energy needs and growing interest in the

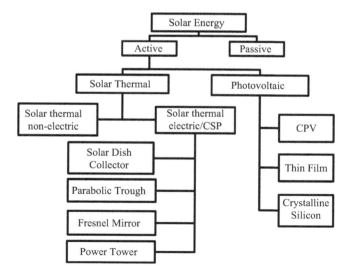

Figure 5.1 Types of modules available globally.

renewable sector have led towards the inclination in the solar sector keeping in view the crisis in oil in 1970s. Thus, researchers and scientists have started to put more stress to harness solar energy by various methodologies. This has made this thrust area a compulsion as a whole. Considering the region as terrestrial, solar energy is harnessed via either of the thermal routes like dryers and heaters or the PV process. The PV converts the energy from the sun in a direct way without involving interface and excess of devices. Considering the countries in the developing domain, above 80% of the population is basically from rural localities that are dependent on the traditional or say conventional sources of energy in order to suffice their energy demands. For example, for cooking, heating or lighting purposes they are dependent on wood or kerosene, the animal livestock is necessary for the activities engaged in agriculture, and also diesel engines to drive water sources to field for irrigation part. The population harnesses the sources as per the increasing demands thus creating an imbalance in the harmony of the nature. This has led to the alarming threat and contributed to the environmental issues related to global warming, overall global, temperature increase, deforestation and even extinction of resources. Very frequently the present generation is subjected to crisis and the prevailing conditions may even lead to blackouts. Thus, there is a need to step into competitive industry 4.0 revolution race but in a healthy and sustainable eco-friendly way. The following figure depicts the energy scenario across the globe.

Hence, there is a tremendous requirement for the sources of energy that are economically as well as environmentally sound and hand in hand also help in developing sustainably [1]. This shall help a lot to satisfy the present energy demands as depicted in Figure 5.2 as well as the future needs also.

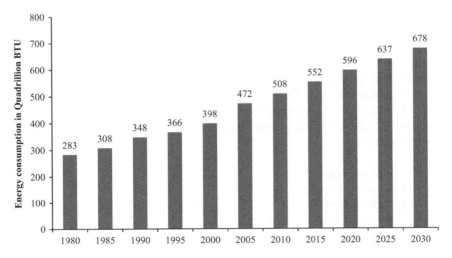

Figure 5.2 Global market energy consumption scenarios [2].

5.2 WHAT IS SOLAR THERMAL?

The regions on the earth receive varied solar irradiance depending on the geographical, latitude and longitude distribution across the globe. The irradiation may have various ranges of intensity and that shall affect the reliability of the same. The technology of solar thermal utilizes the sun's energy for heating up the water for usage purposes. Thus the heat from the sun is absorbed and transferred for heating water for the needful usage purposes like residential, industrial applications, etc. Thus we observe that it is a free-to-use energy from nature with direct and hazel-free application. The heat is transferred from the collectors to the cylinder of hot water. This is the use of thermal energy being converted from the radiations of the sun. The basic difference between this system and PV is that the former converts the radiation to thermal form and the latter to electricity. The broad advantage not only has less maintenance but also a drastically lowered footprint of carbon that overall leads to the cost saving in electricity bills and hence huge savings. The reasons to install thermal hence clearly include the following:

- Meeting the summer demands of hot water to about 90%–100%.
- Minimized greenhouse gas emissions.
- Lowered dependency on fossil fuels which may be very expensive in future.
- Long-term savings in terms of finance.
- Benefit related to the market and public domain.

5.3 PRINCIPLE OF WORKING OF SOLAR THERMAL SYSTEM

The necessary components in the system of thermal type for solar include collectors and a tank of hot water. The collectors of solar work like the panels of solar that are installed on rooftops. This collector absorbs the heat from the sun in a fluid (either water or ethylene glycol or a combination of both) and transfers it to heat up the water via the same fluid to transfer the absorbed heat. This leads to the collector configuration of the following types:

a. Flat-plate type
b. Evacuated-tube type
c. Unglazed type
d. Concentrating solar system
e. Transpired solar air collectors

Advancements in Solar PV System 63

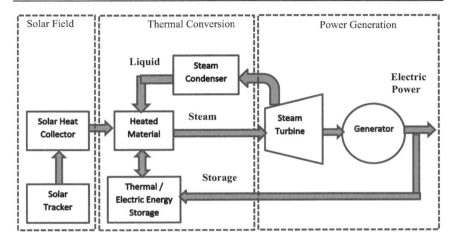

Figure 5.3 Configurational setup of concentrated solar power system.

The preference of the collector is solely dependent on the system requirements including the following factors:

a. Rooftop of the structure
b. Economic aspects
c. Location and weather conditions
d. Type of design required for the system.

The heated water may be for normal home utilization or for boilers as a CSP as shown in Figure 5.3.

The solar thermal works on a normal principle by absorbing heat from the sun with the employment of the solar collectors connected to the solar panels. This leads to the two configurations of the solar thermal system as follows:

a. Active system
b. Passive system.

The active configuration is more expensive but better in operation than the passive configuration. These are mostly employed in domestic hot water (DHW), space heating systems and hot tub heating and swimming pool. Solar is a renewable source as the sun continuously supplies it to the earth's surface.

5.4 WHAT IS SOLAR PV?

The rugged and simply designed modular solar PV system has lesser maintenance requirements. The stand-alone type system is capable of

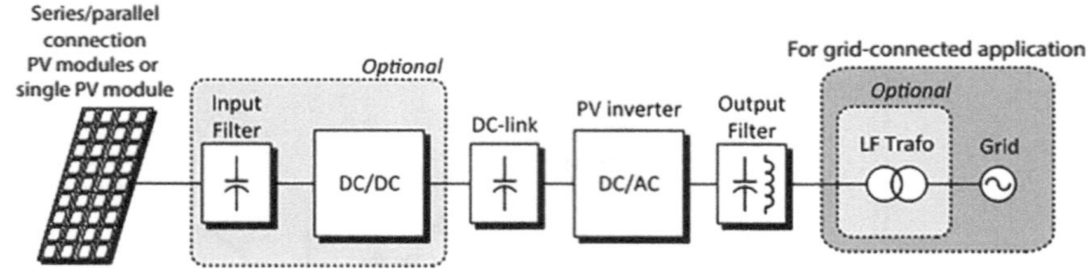

64   Thermal Energy Systems

Figure 5.4 Configurational setup of grid-connected PV system [3].

generating energy ranging from a few microwatts to several megawatts. In countries like India, the rural electrification has been tremendously being supported by the stand-alone PV systems. The very popular Solar Homes can be easily installed with the employment of charge controllers and batteries as per the specifications of the load. This has been reflected in the greater demand for this type. Also, the decline in its cost has led to the popularity of the solar PV systems (Figure 5.4).

The effective policies of tariff by the government also have caused a rise in demand for the PV system installations in developing countries. Thus the comparison of the system variables is done with the static fixed system to prove effective results in terms of reliability, efficiency and cost. As per the figures obtained in a survey till the execution of the year, the overall generation in terms of solar capacity was nearly 629 GW that demonstrated a hike of about 48% as compared to that in 2016.

5.5 RECENT ADVANCES IN SOLAR PV AND THERMAL TECHNOLOGIES

The researchers hold a great interest to explore the systems related to solar harnessing technologies including solar desalination, solar dryers or heaters, solar homes and many more ranging till BIPV (building-integrated PV) too. The work process involves exegetic approaches in the cell used for solar technology for the reason of being a mature and highly efficient one [4–6]. Since its inception, silicon-based ones lead the market. The continuous research followed by wider domains of exploration leads towards looking for alternative solutions for the same cause in order to enhance the efficiency to the utmost. Thus new ways of electricity production is a specific thrust area of research and development these days. Specifically, the authors pose interest in the recent developments in these areas along with the applications of the same. The literature in this domain concludes that the thin film technology and the lowered cost with flexibility make it a perfect market fit. The first advancement shows the cost reduction to 50% of the crystalline topology but it affected adversely the conversion and hence the efficiency part. This was a matter of serious concern. Hence

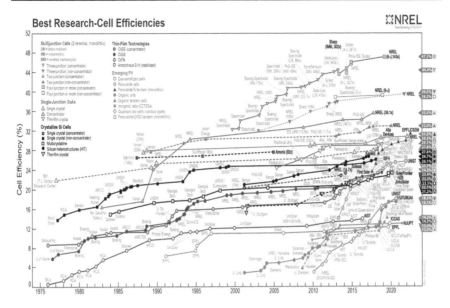

Figure 5.5 Laboratories NRE. Best research cell efficiencies [9].

researchers now are focusing on the materials including CdS or CdTE and also amorphous silicon. Developments are also going on in the CIS for increasing the efficacy of the thin film technology related to PV [7]. The main focus of this technology works on the polymer or organic material keeping in view the concerns related to the environment [8]. Hence increased developments are going on in the field on materials required for cell fabrication. This is summarized in the following figure reflecting the current scenario in the R&D (Figure 5.5).

5.5.1 Recent Advancements in the Solar Cell Materials

Figure 5.6 shows the various solar cell materials present in the market these days. Till date, the silicon-based technology has been given priority all over the globe due to its higher efficiency and unmatched technology. Also, the world is now focusing to increase their production in the environmental concern aspects.

China has taken a lead since many years in the production of PV being followed by Taiwan and Japan as early as in 2013 [10]. Semiconductors headed under crystalline like GaAs and Si reflect superb performance characteristics compared to the others in the market. If we go for the cost reduction we find that options like impure polycrystalline or organic or even inorganic amorphous or a combination of the above-mentioned give results but with lowered efficiencies [11]. So the challenge for the innovators in this field is to work on the higher performance and efficient materials

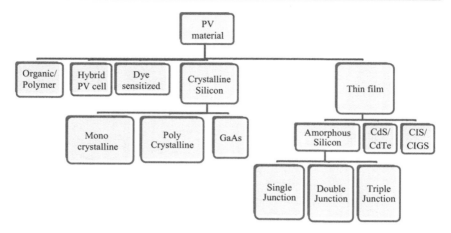

Figure 5.6 Solar PV materials.

but at the same time have a reasonable cost. Also, stress is being laid to develop materials having lesser weight as compared to the poly or mono configurations already prevailing in the market because the name d suggests thin film technology. The experiments are being continuously carried out on the materials like CdS/CdTe and amorphous Si but environmental concerns being the highest threat in the development of the same. The research has been classified in each particular domain as in the following sections.

5.5.1.1 Crystalline Configuration

The configuration is loaded with numerous advantages like high efficiency and very easily available. Monocrystalline is preferred due to its high efficiency but at the same time has cost constraints. Thus recent advances are being made to discover new materials that are suitable for both the manufacturer as well as the end user. Polycrystalline is gaining interest from the experimentalists as it is lower in cost than the monotype. The advance in this field has focused on a solar cell based on GaAs (a compound of Gallium and Arsenic) so as to match the structure of the silicon. This material is higher heat resistance (required for CPS – applicable in hybrid and space applications), lighter in weight, higher in efficiency but also higher in cost as compared to mono or polyconfigurations.

5.5.1.2 Thin Film Technology

Due to the lesser need of manufacturing material for the thin film configuration is available at cheaper rates than the silicon ones, and hence gains popularity and interest these days. The disordered structure and non-crystalline nature of amorphous silicon make it 40 times more light absorptive than the monocrystalline structure. Thus CdS/CdTe and CIS/CIGS are less efficient as compared to the amorphous type. The recent advances include the RF sputtering the ZnO/CDs, etc., with the close space sublimation for growing the configurational substrate. The two-stage process helped to procure the annealing of $CdCl_2$. The first step includes the doping of CdTe and the next step includes the analysis with mass spectrometry in secondary ions by inter-diffusing CdTe/CdS. Thus the approximate 8% increase in efficiency is proved with the involvement of a layer of zinc oxide between the two layers of ITO and CdS. This happens with the increment of the shunt resistance earlier being 563 Ωcm^2 and later being increased to 881 Ωcm^2 (Figure 5.7).

Further, the experimental analysis was carried out by Soliman et al. [13] explaining the need for heat and chemical treatment in order to enhance the efficiency of the type. The studies also show that the lab efficiency of CIGS is approximately 20.3% [14,15]. Further, the annealing and selenization at 450°C in CIGS technology was studied in detail by Jun-Feng et al. [16]. The major challenging aspect is the activation of the junction interfacial and the lead to back contacts being stable. Also, the grain boundary is found with the diffusion of the impurity is causing hindrance in the further progress.

5.5.1.3 Solar PV Concentrated Technology

As the name suggests, here we utilize the solar power by concentrating the sun rays and then harnessing the irradiance. In this technique, the advances focus on developing the minimum cost lenses or mirrors for concentrating the rays to replace the highly expensive solar cells. Researchers also focused on the usage of small size and reduced number of systems with high intense radiations. The major work being carried out in this domain with the life cycle and durability of this system as high intensity may cause the risk in the

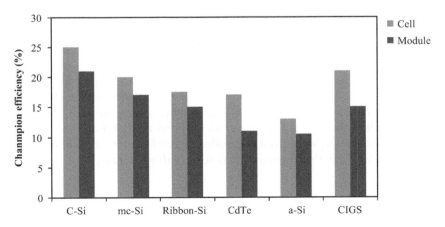

Figure 5.7 Reported champion efficiencies [12].

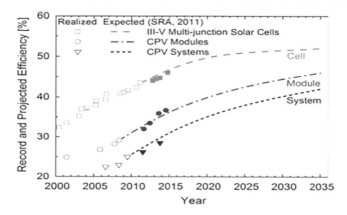

Figure 5.8 Record efficiency of multijunction and CPV modules [19].

construction design part and may damage the cells of the panel. If proper arrangements of cooling are made to utilize the extracted heat energy from the CPV the efficiency may increase to some extent. The experimentation also includes the study related to number of concentrated solar cells including 1 time to 950 times count between 25°C and 85°C [17]. The study reflected that the ratio of dV_{OC} to dT can be reduced and further made constant by elevating the levels of concentration. Even models have been successfully implemented to demonstrate positive results and support energy conservation [18] (Figure 5.8).

5.5.1.4 Polymer and Organic Configuration

Flexible mechanically, robust, reduced cost and lightweight nature of this type is the main attraction to the configuration. This has proved to be an alternative solution to the solar cells maintaining the sustainable aspects. The maximum advances are observed in the domain of developing using a donor poly P3HT and acceptor as (PCMB) Phenyl C60 butyric acid methyl ester including a lot of (BHJ) heterojunction structures have been founded [20,21]. Including the influence of electrical bias and optical application the outer quantum efficiency of solar cells in dual terminal tandem type was observed to be 16% [22]. Pei et al. [23] reported that in the influence of higher electric field, the efficiency of power conversion of this type of cells may hike approximately above 8%. Capasso et al. [24] proved that if on the top of the FTO glass electrode, we grow a controlled layer of carbon nanotubes (multiwall), MWCNT, the performance was remarkably increased.

5.5.1.5 Hybrid Configuration

This type is the mixture of the organic and inorganic materials to develop a solution that is cheap, has better structure and molecular design and has better efficiency at the same time. The combination of organic material makes the optical absorption strong while the semiconductor inorganic in nature causes the charge carrier mobility to increase. Such a material was designed and tested by Zhang et al. [25] for fill factor (FF) and open circuit voltage V_{OC} with amorphous and pure silicon intrinsic in nature, P-doped hybrid bilayer silicon structure including poly (3-hexylthiophene) and B-doped silicon. The features were found to be moderate and dependent strongly on the molecular Si: H doping films.

5.5.1.6 Dye-Sensitized Type

This type is favoured due to the reason of reduced costs and simpler design as compared to the traditionally followed schemes. In the year 1991, O'regan and Gratzel [26] invented the design of the said type. This configuration follows the generalized principles as follows:

a. TiO_2 as the semiconductor film
b. Redox mediator in electrolyte
c. The semiconductor surface adsorbed by a sensitizer
d. An oxide that is transparent conductive as coating for a mechanical support
e. An electrode as a counter that can regenerate redox mediator, e.g., Platine

Platinum-free electrodes (counter) were discovered by Ahmad et al. [27] to have a power conversion efficiency of 4% approximately a bit higher than Pt base CE for the type of DSSC (dye-sensitized solar cells).

5.6 ADVANCED APPLICATIONS AND TRENDS IN SOLAR PV

For the concern related to the future needs and a clean environment, the scientific methods sustainable in nature are being studied and developed continuously. Also, the fast depletion of the conventional sources is a major concern triggering the advancements. The solutions are found to be competitive, healthy and sustainable keeping the economic growth in the focus. The maximum advances are carried out in the thrust areas like PV-based home lighting, desalination plants, water pumping, thermals, integrated buildings, space technologies and CSPs. The developments are observed globally in all domains let it be public, government or private type. This has speeded up the real-life applications in the renewable sector. The growing incentives are also a major attraction. The studies reflect that there is a need to enhance the reliability, power quality and efficiency. At the same time, the focus is on lowering the cost and integration with the grid, especially for the paradigms of smart-grid and micro-grid sectors. The developments enlisted promise a sustainable solution if worked upon with the factors rigorously. The drastic change maker is the novel material technology for multi-junction configuration of solar cells. The converter efficiency and fault-tolerable operations having reduced overall costs (including transportation and installation) with systems integrated into multilevel inverter topologies have made a remarkable change. The multilevel inverter with increased levels of voltage involving solar power at the input has proved to enhance the power quality with reduced THD (total harmonic distortion) [28–30]. The suggested converter topology for 15-level MLI with appropriate PWM reduces the THD to 3.17% as compared to the conventional model. The simulation also projected the cost reduction due to the involvement of less switches and hence low switching losses. Finally, it must be noted that the growth towards a clean energy source may be obtained by the joint efforts of the Governments and the potential population around the globe [31–34].

5.7 CONCLUSION

Solar technology systems have proved to be undoubtedly the most promising solutions all around the world. The chapter has summarized the prominent and considerable advances in this field. Many revolutionary changes regarding solar materials are still under investigation and testing phases. This profitable investment is found to be compact, more efficient and cheap. Though the systems have proved to be a great boon, still there are major challenges like design, weather and geographical constraints that need to be addressed with competent and proven technologies.

REFERENCES

1. Patra, B.B. Smart grid-sustainable shaping of the future smarter nation. *Int J Emerg Technol Adv Engineer, First Int Conf Innov & Eng* 8. 2018.
2. Türkay, B.E., Telli, A.Y. Economic analysis of standalone and grid connected hybrid energy systems. *Renew Energy* 2011; 36: 1931–1943.
3. Kouro, S., Leon, J.I., Vinnikov, D., Franquelo, L.G. Grid-connected photovoltaic systems: An overview of recent research and emerging PV converter technology. *IEEE Industrial Electron Magazine* March 2015; 9(1): 47–61.
4. Pandey, A., Tyagi, V., Tyagi, S. Exergetic analysis and parametric study of multicrystalline solar photovoltaic system at a typical climatic zone. *Clean Technol Environ Policy* 2013; 15: 333–343.
5. Pandey, A.K., Pant, P.C., Sastry, O.S., Kumar, A., Tyagi, S.K. Energy and energy performance evaluation of a typical solar photovoltaic module. *Therm Sci* 2013; 147: 147.
6. Elhadidy, M. Performance evaluation of hybrid (wind/solar/diesel) power systems. *Renew Energy* 2002; 26: 401–413.

7. Meena, C.S., Kumar, A., Roy, S., Cannavale, A., Ghosh, A. Review on boiling heat transfer enhancement techniques. *Energies* 2022; 15: 5759. doi: 10.3390/en15155759
8. Gorter, T., Reinders, A. A comparison of 15 polymers for application in photovoltaic modules in PV-powered boats. *Appl Energy* 2012; 92: 286–297.
9. Singh, V.P., Jain, S., Karn, A., Kumar, A., Dwivedi, G., Meena, C.S., Cozzolino, R. Mathematical modeling of efficiency evaluation of double-pass parallel flow solar air heater. *Sustainability* 2022; 14: 10535. doi: 10.3390/su141710535
10. Singh, V.P., Jain, S., Karn, A., Kumar, A., Dwivedi, G., Meena, C.S., Dutt, N., Ghosh, A. Recent developments and advancements in solar air heaters: A detailed review. *Sustainability* 2022; 14: 12149. doi: 10.3390/su141912149
11. Nayak, P.K., Garcia-Belmonte, G., Kahn, A., Bisquert, J., Cahen, D. Photovoltaic efficiency limits and material disorder. *Energy Environ Sci* 2012; 5: 6022–6039.
12. Wolden, C.A., Kurtin, J., Baxter, J.B., Repins, I., Shaheen, S.E., Torvik, J.T., et al. Photovoltaic manufacturing: Present status, future prospects, and research needs. *J Vacuum Sci Technol A* 2011; 29: 030801.
13. Soliman, M.M., Shabana, M.M., Abulfotuh, F. CdS/CdTe solar cell using sputtering technique. *Renew Energy* 1996; 8: 386–389.
14. Contreras, M.A., Romero, M.J., Noufi, R. Characterization of Cu (In, Ga) Seosub42o/sub4 materials used in record performance solar cells. *Thin Solid Films* 2006; 511: 51–54.
15. Singh, V.P., Jain, S., Kumar, A. Establishment of correlations for the thermo-hydraulic parameters due to perforation in a multi-V rib roughened single pass solar air heater. *Experimental Heat Transfer* 2022. doi: 10.1080/0891 6152.2022.2064940
16. Han J-f Liao, C., Jiang, T., Xie, H-m, Zhao, K. Investigation of Cu (In,Ga) Se$_2$ polycrystalline growth: Ga diffusion and surface morphology evolution. *Mater Res Bull* 2014; 49: 187–192.
17. Cotal, H., Sherif, R. Temperature dependence of the IV parameters from triple junction GaInP/InGaAs/Ge concentrator solar cells. *Photovolt Energy Convers* 2006: 845–848 (IEEE conference on record of the 2006 IEEE 4th World Conference).
18. Kuo, C-T, Shin, H-Y, Hong, H-F, Wu, C-H, Lee, C-D, Lung, I., et al. Development of the high concentration III–V photovoltaic system at INER, Taiwan. *Renew Energy* 2009; 34: 1931–1933.
19. Singh, V.P., Jain, S., Karn, A., Dwivedi, G., Alam, T., Kumar, A. Experimental assessment of variation in open area ratio on thermohydraulic performance of parallel flow solar air heater. *Arab J Sci Engineer* 2022, 10. 1007/s13369-022-07
20. Bagienski, W., Gupta, M. Temperature dependence of polymer/fullerene organic solar cells. *Sol Energy Mater Sol Cells* 2011; 95: 933–941.
21. Parvathy Devi, B., Wu, K-C, Pei, Z. Gold nanomesh induced surface plasmon for photocurrent enhancement in a polymer solar cell. *Sol Energy Mater Sol Cells* 2011; 95: 2102–2106.
22. Gilot, J., Wienk, M.M., Janssen, R.A. Measuring the external quantum efficiency of two terminal polymer tandem solar cells. *Adv Funct Mater* 2010; 20: 3904–3911.

23. Pei, Z., Parvathy Devi, B., Thiyagu, S. Study on the Al–P3HT: PCBM interfaces in electrical stressed polymer solar cell by X-ray photoelectron spectroscopy. *Sol Energy Mater Sol Cells* 2014; 123: 1–6.
24. Capasso, A., Salamandra, L., Chou, A., Di Carlo, A., Motta, N. Multi-wall carbon nanotube coating of fluorine-doped tin oxide as an electrode surface modifier for polymer solar cells. *Sol Energy Mater Sol Cells* 2014; 122: 297–302.
25. Zhang, Z., He, Z., Liang, C., Lind, A.H., Diyaf, A., Peng, Y., et al. A preliminary development in hybrid a-silicon/polymer solar cells. *Renew Energy* 2014; 63: 145–152.
26. O'regan, B., Gratzel, M. A low-cost, high-efficiency solar cell based on dyesensitized colloidal TiO_2 films. *Nature* 1991; 353: 737–740.
27. Ahmad, I., McCarthy, J.E., Bari, M., Gun'ko, Y.K. Carbon nanomaterial based counter electrodes for dye sensitized solar cells. *Sol Energy* 2014; 102: 152–161.
28. Patra, B., Nema, P. Comparative design analysis of modified solar based 15 level multi level inverter for power quality improvement. *J Algebraic Stat* 2022; 13(3); 1084–1095.
29. Patra, B.B., Nema, P. Novel topological analysis of 27 level three phase multilevel inverter. *NeuroQuantology Scopus Index J* 2022; 20(14): 6, ISSN: 1303 5150, doi: 10.4704/nq.2022.20.14.NQ88052
30. Patra, B., Nema, P., Khan, M.Z., Khan, O. Optimization of solar energy using MPPT techniques and industry 4.0 modelling. *Sustain Operations Comput, Scopus Indexed J* 2022, ISSN:2666-4127, doi: 10.1016/j.susoc. 2022.10.001
31. Singh, V.P., Jain, S., Karn, A., Dwivedi, G., Kumar, A., Mishra, S., Sharma, N.K., Bajaj, M., Zawbaa, H.M., Kamel, S. Heat transfer and friction factor correlations development for double pass solar air heater artificially roughened with perforated multi-V ribs. *Case Studies Thermal Engineer* 2022; 39: 102461, ISSN 2214-157X, doi: 10.1016/j.csite.2022.102461
32. Meena, C.S., Kumar, A., Jain, S., Rehman, A.U., Mishra, S., Sharma, N.K., Bajaj, M., Shafiq, M., Eldin, E.T. Innovation in green building sector for sustainable future. *Energies* 2022; 15: 6631. doi: 10.3390/en15186631
33. Dutt, N., Binjola, A., Hedau, A.J., Kumar, A., Singh, V.P., Meena, C.S. Comparison of CFD results of smooth air duct with experimental and available equations in literature. *Int J Energy Resour Appl (IJERA)* 2022; 1(1): 40–47. doi: 10.56896/IJERA.2022.1.1.006
34. Kushwaha, P.K., Sharma, N.K., Kumar, A., Meena, C.S. Recent advancements in augmentation of solar water heaters using nanocomposites with PCM: Past, present & future. *Buildings* 2022;13: 79. 10.3390/buildings13010079

Chapter 6

Thermal Energy Applications in Net-Zero Energy Buildings

Mrityunjai Verma and Ashish Karn

CONTENTS

6.1 Introduction 74
6.2 Thermal Comfort and Air Quality Requirements in nZEB 76
6.3 Heat Transfer Mechanisms in Building Structures 78
      6.3.1 Strategies to Regulate Heat Transmission in Building
            Structures 78
            6.3.1.1 Homogeneous Structures.. 79
            6.3.1.2 Structures with Closed-Air Gap or
                    Ventilated Air Layer.. 79
            6.3.1.3 Ground-Contact Building Structures 81
            6.3.1.4 Windows and Doors 81
6.4 Heat Generation Technologies for nZEB 81
6.5 Decentralized Heat Generators 83
      6.5.1 Biomass Stoves and Furnaces 83
      6.5.2 Electrical Heaters 83
6.6 Heat Generators for Central Heating Systems 84
      6.6.1 Combustion Boilers... 84
            6.6.1.1 Thermal Efficiency of Combustion Boilers 85
      6.6.2 Heat Pumps... 86
      6.6.3 Solar Thermal Collectors 86
      6.6.4 District Heating... 87
6.7 Space Heating of nZEB.. 87
      6.7.1 Building Heat Load.. 87
      6.7.2 Basic Componental Analysis of Space Heating
            Systems and Subsystems... 88
            6.7.2.1 Conceptualization of Heat Containment......... 88
      6.7.3 Distribution Systems 89
      6.7.4 Hydronic Systems.. 89
      6.7.5 End Heat Exchangers.. 90
      6.7.6 Radiators.. 90
      6.7.7 Active Beams... 90

DOI: 10.1201/9781003395768-6

74 Thermal Energy Systems

6.8   Thermally Activated Building Structures..91
      6.8.1   Space Heating System: Control....................................91
      6.8.2   Domestic Hot Water Heating (DHW) in nZEB............91
      6.8.3   Space Setup Cooling Conceptualization for nZEB
            Structures..92
      6.8.4   Mechanical Cooling Systems and Concepts for nZEB ...92
      6.8.5   Direct Evaporation Based Air Conditioning Systems.....93
6.9   Ventilation Systems and Management in nZEB......................93
      6.9.1   Natural Ventilation..93
            6.9.1.1   Working of Natural Ventilation Subsystems...93
      6.9.2   Mechanical Ventilation..95
      6.9.3   Mechanical Ventilation Systems..................................96
      6.9.4   Heat Recovery Units (HRU) and Air Handling Units
            (AHU)..96
6.10   Technological Aspects of the Improvement of
Ventilation Energy Efficiency..98
      6.10.1   Increment in Thermal Recovery-Based Prospects by the
            Utilization of a Ground Heat Exchanger.....................98
      6.10.2   Increasing Energy Efficiency of Buildings With
            Integration of Building Service Systems........................98
References..100

6.1 INTRODUCTION

With the onset of technological development, the world is also experiencing humongous amounts of pollution levels in different forms that are proving to be hazardous for the living population in various ways. One such segment of technology consumption comes under the energy sector. Energy consumption is one of the very basic needs of every human being living on the planet and hence it drives one of the largest consumption channels in the human lifestyle. With the enlargement of the energy consumption sector, what also increases is the emission levels that are associated with carbon footprints. This turns out to be a consequence of increased CFC emission-based energy consumption, heating and refrigeration systems and even basic amenities like electricity. A net-zero energy building, or nZEB, is often a grid-connected structure that has exceptional energy performance. nZEB ensures that its consumption of primary energy is in equilibrium by ensuring that the amount of primary energy that is sent into the grid or any other energy network is equal to the amount of primary energy that is supplied to nZEB via energy networks. When the annual balance of primary energy usage is 0 kWh/(m² a), the circumstance often results in a large percentage of the onsite energy production being traded with the grid.

Therefore, a net-zero energy building will only create energy when the circumstances are optimal and will rely on provided energy during the other times. A significant reduction in energy consumption and carbon emissions must be mandated for nearly zero energy buildings (nZEBs) if high efficiency in buildings is to be achieved via the use of indicators that are widely accepted and thoroughly detailed. This would in no way restrict the potential of tailoring the aims and thresholds level of those indicators to the particulars of the local environment. On the other hand, it would make it possible to have a standard language throughout Europe, which is very necessary for the building sector in order to design solutions inside a framework that is consistent and unified. This is very important because beginning in 2020, all newly constructed buildings will be required to exhibit excellent energy performance, and a considerable portion of their decreased or ultra-low-energy consumption will be provided by sources of renewable energy. The calculation of energy demands for heating, cooling and hot water as well as energy consumption for lighting and ventilation is required for the evaluation of the energy efficiency of novel building structures and old technology-based structures to existing buildings. A significant reduction in energy consumption and carbon emissions must be mandated for nearly zero energy buildings (nZEB) if high efficiency in buildings is to be achieved via the use of indicators that are widely accepted and thoroughly detailed. This would in no way restrict the potential of tailoring the aims and thresholds grade of those metrics to the particulars of the local environment.

On the other hand, it would make it possible to have a standard language throughout Europe, which is very necessary for the building sector in order to design solutions inside a framework that is consistent and unified. This is very important because beginning in 2020, all newly constructed buildings will be required to exhibit excellent energy performance, and a considerable portion of their decreased or ultra-low-energy consumption will be provided by sources of renewable energy. The calculation of energy demands for heating, cooling and hot water as well as energy consumption for lighting and ventilation is required for the evaluation of the energy efficiency of new buildings and retrofits to existing buildings (Medved 2019).

The thermal applications of nZEBs are discussed in this chapter from a variety of different points of view. This chapter contains a comprehensive discussion on the many different approaches that may be used to quicken the pace at which the idea of nZEBs can be implemented. This encompasses the systems and processes for heating, refrigeration and air conditioning, as well as energy-efficient ventilation systems, with the end goal of reaching a net-zero energy consumption level for building structures. In addition to this, it provides a comprehensive study of nZEBs based on heat transfer processes and thermal sciences involved with the development and installation of devices and subsystems that are intended to serve as the foundation for the production of nZEBs. The chapter presents the best available technologies (BAT) for nZEB subsystems and setups, which are either in the stage of research and development or are

76 Thermal Energy Systems

being actively used in contemporary building structures. These technologies are either in the research and development stage or are actively being used in contemporary building structures. There will be a wealth of opportunities for research and development on self-sufficient buildings when the future prospects of nZEBs being put into applications are realized. These buildings would be able to create process and re-process exhaust and waste without having any contact with energy supplies that are obtained from the outside.

6.2 THERMAL COMFORT AND AIR QUALITY REQUIREMENTS IN nZEB

Within the biosphere of living organisms, human beings are categorized as being warm blooded. The human body constantly provides a temperature ranging 37°C ± 0.8°C, regardless of the ambient temperature or activity of the muscles. As a result of the combustion of protein and fats, the body organs produce heat. This is known as metabolism. Hence, similar to the working of power plants, the surplus energy has to be transferred to the surroundings in order to maintain a state of equilibrium, using various mechanisms of heat transfer. If the exchange of heat to or from the body is within the scales of being comfortable, it can be considered to be the achievement of thermal comfort (Medved 2019).

In 1982, Fanger et al. executed an analysis by comparing the experiences of users based on the impact of heat transfer procedures in the human body. Hence, a method to combine the effect of different physical parameters based on the overall thermal comfort. This is calculated in the form of a so-called predicted mean value (PMV). The scale of the PMV according to the research is placed between a cold and a hot mark. This research was performed in order to test the levels of thermal comfort that is considered to be optimal and under the categorization of equilibrium. The percentage of users who were dissatisfied with the thermal comfort levels was categorized under the predicted percentage of dissatisfied or PPD, also proposed within the research (Figure 6.1).

The measures given are known as global indoor thermal comfort indices. As previously mentioned, they are measured at the living zone's centre and used in the design of structures and building services. To address a particular region of the room, such as the working space in a large office, indicators such as the vertical temperature gradient, radiant temperature asymmetry, floor temperature and draught rating are used. Acceptable value ranges for each of these indicators are set depending on the kind of indoor environment and its impact on PMV and PPD values (Figure 6.2).

Inadequate indoor environments can be a major cause of the occurrence of major and minor health problems. Poor indoor air quality results in the occurrence of the following two phenomena:

Thermal Energy Applications in nZEBs 77

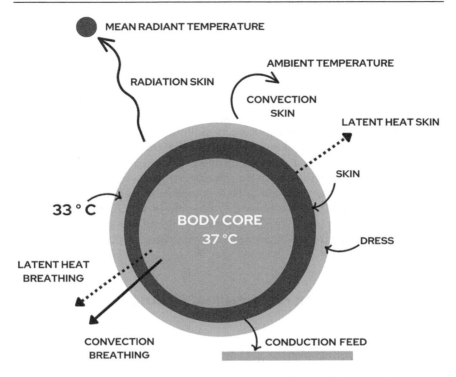

Figure 6.1 A schematic diagram explaining the various forms of heat transfer process execution relevant to indoor thermal comfort (Fabbri 2015).

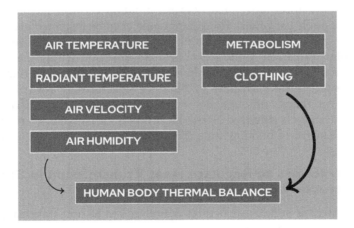

Figure 6.2 Six defined parameters are utilized in order to assess indoor thermal comfort. The figure depicts the correlation of the parameters with the thermal balance of the human body.

78 Thermal Energy Systems

- Sick building syndrome (SBS): SBS is a condition that causes headaches, concentration issues, dry throat, mucous membrane irritation and nausea. People are affected by SBS when inside the structure. The specific aetiology of SBS is unknown, and symptoms fade quickly when the affected individuals leave the premises.
- Building-related illness (BRI): A number of illnesses, such as lung infections, viral diseases and bacterial diseases, manifest themselves as a result of prolonged exposure to pollutants in the building. Cancer and cardiovascular illness have also been related to indoor air pollution. BRIs imply a permanent decrease in health. A common technique for replacing the stale interior air with fresh outdoor air is ventilation. Often, the air outside is cleaner than the air inside. Both natural and forced ventilation is possible in buildings (mechanical). Air flow is created by the difference in air pressure between the interior and exterior environments, regardless of the type of ventilation used (Lenz et al. 2012).

6.3 HEAT TRANSFER MECHANISMS IN BUILDING STRUCTURES

Building constructions typically entail all three modes of heat transfer: conduction heat flux through the solid layers giving way to a convective heat flux and radiation heat transfer. Moving air molecules create convection heat flux, while walls exchange thermal radiation with the environment, sky, and interior surfaces and heat sources. Convection and long-wave radiation move heat between building surfaces having a closed gap with a noble gas or air. These gaps could even be insulated glazing, or a vented air gap as found in ventilated roofs. Exterior surfaces absorb shortwave (solar) radiation. Conduction heat transmission within the structure resulted in sensible or latent heat buildup if phase change materials (PCM) are employed. Heat transport in building structures is always dynamic (unsteady in time), although it's a steady method for identifying basic thermal parameters (Medved 2014). Heat flux is measured in watts or watts per square metre of building area (Figure 6.3).

6.3.1 Strategies to Regulate Heat Transmission in Building Structures

Heat transfer for building structures is analyzed to be non-stationary due to it always being time-dependent. Nonetheless, determining heat transfer under stationary settings is sufficient for the initial verification of a building envelope's thermal insulation qualities. Thermal transmittance of a building is employed for this purpose. In terms of heat transmission, buildings may be classified as homogenous structures with closed gas-filled spaces,

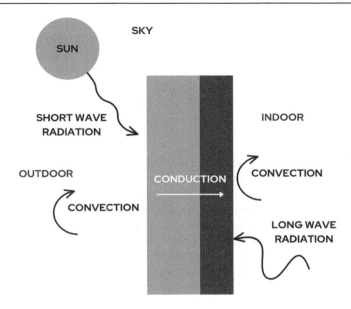

Figure 6.3 Building structures use heat fluxes based on conduction, convection and radiation for heat transfer at different stages (Flanders 1994).

structures with open spaces (that may allow for natural or forced), greened, and structures having ground contact (Asdrubali 2014).

6.3.1.1 Homogeneous Structures

These constructions are made of homogenous construction materials, which are unaffected by temperature or humidity and have the same physical properties throughout the volume. The heat flux is in a direction perpendicular to the surface of the structure, and the external and inner surfaces, as well as all layers, are parallel. Given that brick walls are adhered with mortar and that the heat conductivity of fibrous insulating materials depends on the orientation of the fibres, for example, it is simple to argue that nearly no buildings fit the requirements of this description. However, because these anomalies barely affect heat transport, the majority of building constructions can be regarded as homogenous. In this instance, the heat transmission is approximated by figuring out the thermal resistances between the temperature nodes.

6.3.1.2 Structures with Closed-Air Gap or Ventilated Air Layer

Air gaps in building structures are extremely prevalent. A closed or ventilated air gap may be built. When the thickness inequalities between building structural layers are levelled, closed-air spaces occur. These closed-air gaps increase resistance to fire and lower thermal transmission. The closed (unventilated) air gap transmits heat through convection and radiation. As a result, the thermal resistance of a closed-air layer and that of an unventilated air gap are combined. Heat flow direction, emissivity and air gap thickness all have an impact on thermal resistance.

Green Structures
Green buildings typically employ plant cover atop the structures that play an important part in managing city microclimates and minimizing heat transmission into buildings, particularly during summer. Plants on top of the buildings operate as mini shades owing to their strong absorptivity of short-wave solar radiation, and also due to the high emissivity in the long-wave thermal radiation range. Thus, they keep cool by the means of radiative cooling such as evapotranspiration, which refers to the evaporation of soil water and water carried by the root system via leaf pores from the grooved substrate (transpiration). In spite of the coupled mass and heat transfer processes, the thermal transmittance is calculated in a manner similar to the homogeneous structures, except that the thermal resistance of the layers touching rainfall is not considered. As a result, the rainproofing layer is the last layer exposed outside and its thermal resistance is taken into account. Green constructions, like all other building structures, must have a lower U-value than the permissible value (Asdrubali 2014) (Figure 6.4).

Figure 6.4 A picture of a typical green building where the heat transfer to building interiors is minimized (Sharma 2020).

6.3.1.3 Ground-Contact Building Structures

Since the ground temperature is typically higher, it undergoes less fluctuations than the ambient air and also because the ground is an important contributor to the thermal resistance from the buildings, heat transfer scenario in ground-contact buildings is different from other buildings. Hence, the ratio of the ground-contact area in construction to construction circumference is an important parameter that has a significant impact on heat transport. The techniques for calculating thermal transmittance of building structures in ground-contact varied depending on the kind. These are classified as horizontal buildings with an insulation layer and vertical structures erected partly or entirely under the ground (Medved 2002).

6.3.1.4 Windows and Doors

In order to reduce the heat transmission through windows and doors, several steps may be taken such as filling the glazing gaps in the windows with noble gases such as Krypton, that are denser and have lesser thermal conductivity. Also, low-emission coatings are applied to the glass, which considerably reduces radiation heat transmission between the glass panes. Multiple glass panes may also reduce the thermal transmittance of window glazing. The marker has glazing with up to four panes. Similarly, for frames made of plastic or metal, the thermal transmittance can be diminished by breaking the thermal bridge produced on the frame, realized physically by dividing the frame into separate sections. Typically, windows suitable for installation in nZEBs or energy-efficient buildings have frames that are wrapped with insulation layers or closed-air spaces (Figure 6.5).

The optical properties of the glass undoubtedly affect how the solar radiation passes through window glazing and some other transparent materials on the exterior of the building. While part of the incoming solar energy is absorbed by the window panes, some are reflected back into the environment. As a result, greater heat flux is transferred into the interior space by convection and long-wave radiation, and the temperature of the interior glazed face may be higher than that of the indoor air. The ratio of the total heat flux transported to the indoor environment to the solar irradiance on the exterior surface is referred to as the overall solar energy transmittance or g-value (Medved 2014).

However, apart from these heat transfer considerations in different kinds of structures, nZEBs also entail other thermal factors such as heating, cooling and ventilation requirements in order to meet the criterion for thermal comfort and the desired air quality.

6.4 HEAT GENERATION TECHNOLOGIES FOR nZEB

Decentralized or centralized modes can be used to generate heat for nZEB. A conventional local heater distributes heat evenly throughout a single area,

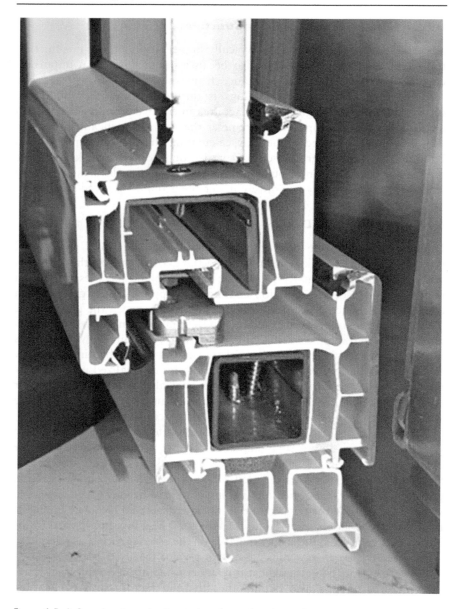

Figure 6.5 A five-chamber plastic window frame for thermal transmittance operations in building structures (Vallabhy 2019).

and the room's temperature is controlled from a single-point source using the convection and radiation modes. In contrast, a central heating system distributes heat from a heat generator to a heat transfer fluid (such as air, water or other), which then travels through a heat distribution pipe to a room heat emitter. A number of heat generators could be added to a single heating system to allow for the use of diverse energy sources throughout the year, to offer backup heating (like solar heating or heat pumps), or to adjust heating power to the heating load through parallel operation. The central heating system requires more complex control to make sure that the air temperature is the same as (or close to) the set-point temperature, boosting thermal interior comfort and energy efficiency. In addition to often being more cost-effective than distributed heat sources, well-managed central heating systems often require less maintenance and are more adaptive to changing energy sources or utilizing a variety of energy sources (Medved 2019).

6.5 DECENTRALIZED HEAT GENERATORS

6.5.1 Biomass Stoves and Furnaces

Since they draw oxygen directly from the air in the room, outdoor fireplaces possess reduced thermal efficiency (less than 50%) and it may result in poor indoor air quality (IAQ). Poor ambient air quality is a problem in many nations, particularly in urban areas. Log burning in cold areas is made possible by the thermal efficiency and heat buildup of clay brick burners with hot air ducts. Low-energy buildings frequently use clay brick stoves with superior thermal mass and closed-combustion biomass stoves. Biomass is a renewable energy source that can be produced locally and doesn't release any CO_2. Modern biomass burners must have a blower to remove flue gases and an outdoor (combustion) air intake conduit with a minimum diameter of 80 mm. This lowers pollutants and boosts thermal efficiency (by 90%). Modern appliances have improved mechanical components, control systems and prefabricated wood fuels like pellets for high thermal efficiency, low environmental emissions and pleasing operation. Unburned particulates are transferred directly to inbuilt ash storage while combustion air is delivered by a fan. A considerable temperature differential in the interior air is produced by substantial thermal power and vigorous convection heat transfer, which lowers heating efficiency and operating control. High surface temperatures could make working conditions uncomfortable. This component is used in conjunction with a centralized heat generator to heat during warm weather (Nussbaumer 2008) (Figure 6.6).

6.5.2 Electrical Heaters

Due to large harmful gases and the high overall resources consumed for electricity generation, electrical heaters are not particularly suited for persistent heating in low-consumption buildings. Electrical lights, on the other extreme, should always be considered if thermal demand seems to be unusually low or whether interior air temperature should be improved

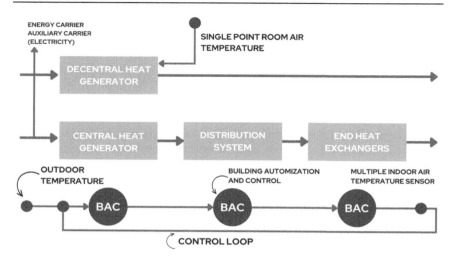

Figure 6.6 Single-point room thermostats operate decentralized heat generators, which are commonly used for separate space heating. On the other hand, central heating systems have more end-heat emitters and distribution systems. These systems are managed by advanced construction control and automation systems, resulting in improved thermal comfort and increased energy efficiency (Medved 2019).

by boosting the radiant temp. Electrical responsive heaters (also known as Watt lamps or ohmic radiators) produce heat in proportion to a resistance value and the quadratic of the current. Such heaters can be passive buildings as built-in heating booster packs, or also may be incorporated into the ventilation system by allowing overall warm air heating, and can be cost-efficient in detached constructs or designed high-temperature foils (Figure 6.7).

6.6 HEAT GENERATORS FOR CENTRAL HEATING SYSTEMS

6.6.1 Combustion Boilers

Fire boilers are frequently used for the heat burners in central heating installations when burning fossil of biomass fire. A detonation boiler consists of both a burner that delivers an acceptable nice mix of fuel air, flywheel of the concoction, a torching compartment inside which heat is released, a temperature-controlled heat exchanger inside which heat is applied to that same thermal performance media, a flues connection for such flow waste heat or a control system (Singh 2014). Fans are utilized in stoves for liquid (home heating, bio-oil, as seen in this image), gasses (NG, Mpg, pure biogas) and combustibles to provide a link found for both fuel and combustion air, and producer gas ambient burners (NG, LPG) (Oland 2002).

Figure 6.7 An illustrative example of underfloor heating. The setup is used to prevent 'cold feet' and to maintain equilibrium on the surface (Veld 2004).

6.6.1.1 *Thermal Efficiency of Combustion Boilers*

The energy performance of burning boilers is impacted by the quality of something like the complete combustion. Energy losses as in flue gas and ambient temperature through to the boiler circuit is huge. These heat losses can be use to heat any building or the era of design and contemporary operation. Centrifugal, ambient, and lava burns all have the same work to provide the optimal combination of air/fuel air. Electrical/Electronics burners can ignite the mixing of liquid or solid food or or traditional biomass (Medved 2019). Water heating ignition boilers are classified based on their working temperature:

> Exhaust gas temperatures in strong boilers must be always better than the dew-point temperature of moisture into flue gases over operation and eliminate water vapour condensation.

Low-temperature boilers: If such a boiler is made of condensation-hardened steel, excess water condensation is allowed occasionally; the temperature of the heating water may be reduced (45–65°C), resulting in less heat management through flue gases (Kum 2013).

6.6.2 Heat Pumps

Buildings have become much more energy efficient in recent decades. As a consequence, the heat load and energy requirement of warming every unit of improvement of structures are substantially lower than those in 'pre-EPBD' constructions. In addition, a lower tier thermodynamic efficiency of the suitable temperatures indicates the lower temperature for the transfer fluid. This is why heat producers referred to as pumps (HP) have recently gained popularity. Heat pumps generate usable heat by transferring the energy of renewable heat sources such as air from the atmosphere, soil or underground water into heat and power.

Despite having the lowest efficiency, air-heat pumps serve as the most often used heating element in heating systems. This really is mainly due to the sudden decline in capacity and effectiveness as ambient air temperatures decrease, the relatively high-temperature difference inside the exchanger, as well as the energy needed for occasional refreezing as well as trying to operate this same fan, that also tends to boost the atmospheric flow of air through the evaporator. Frost would therefore collect on the evaporator coil outside at ambient temperatures that range from 0°C to 6°C in moderate and humid conditions, resulting in diminished heat recovery capacity and performance. Defrosting coils is performed by restarting the heat recovery cycle or by other, less effective methods (Arteconi 2018).

6.6.3 Solar Thermal Collectors

Open-loop solar thermal systems frequently employ air solar thermal collectors. Environment air enters these solar collectors in such devices, warms up, and then leaves the building as or before the ventilation system. Another use for ASTC is crop drying. To heat the home and household water, liquid thermal collectors were frequently utilized in closed-loop solar heating systems. This heat transfer fluid is propelled by thermal buoyancy (in heat transfer systems) or a motor as it travels through a circular channel between solar collectors and thermal mass. An LSTC may have a flat surface or a tubular shape, depending on the design of its casing. The components of a solar flat plate heat collector include a shell, a bottom insulating layer, a top clear glass cover, or an absorption with associated pipelines or channels. Solar collectors for atmospheric thermal energy employ an inner glass cover. A little amount of purified water is used as the heat transfer fluid in the heater, which is a copper conduit that is sealed yet under pressure. Water vapour condenses and descends to the bottom of the heat exchanger, where it comes into touch with the absorber, as the heat transfer fluid cools at the top of the pipe. Heat transfer is high as a result of the deformation process, and the working gas does not need to be circulated within (Medved 2019).

6.6.4 District Heating

A large single energy generator as well as a piping-distributed generation constitute district central heating. Buildings are linked to district heating systems by heating systems situated in heat terminals. Flood water is now the most often used heat medium for structure heating, but steam continues to be utilized in industrial settings. The key benefits of the transmission and distribution system include the following:

- The placement of the warming station demands less area than a boiler room.
- In heating buildings, domestic heating stoves are more efficient.
- Lower energy prices on the market may be acquired by more variable manufacturer modifications.
- Flue gas cleaning cleaners are more accurate and valuable due to their larger frame.
- More use of sources of renewable power such as biogas, photovoltaics or solar heat as possible with lower capital costs.
- District heating seems to be particularly cheaper when power and heat are co-generated.
- Due to significant transportation heat, generally smaller district air conditioners only operate during the heating season. In this case, an additional DHW heat producer must be built and kept functioning in during summer (Werner 2017).

6.7 SPACE HEATING OF nZEB

It is possible to develop a local or central system for space heating of nZEBs. These systems are developed by the aid of a stove or instruments like heaters that have a sole purpose of providing warmth within the particular space. The heat is transferred through a heat generator by the means of the process of radiation and convection to the internally existing air. Combined heat and power systems are more extensively used in mild and cold climates because they provide a significant thermal environment and have a higher overall energy efficiency. In this situation, in addition to the heat generator, a heat exchange medium, distribution lines, and end-heat exchangers positioned in the heated region are required (Kabele et al. 2012).

6.7.1 Building Heat Load

The building heat load is utilized to govern the values of the thermal capability of a heat generator when subjected to maximum load. The heat losses relevant to ventilation are responsible for analyzing the heat load at a particular interior space and outdoor air temperature (Medved 2019) (Figure 6.8).

88 Thermal Energy Systems

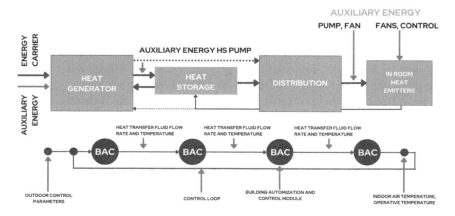

Figure 6.8 A schematic representation of the various subsystems of a hydronic space heating system (Medved 2019).

6.7.2 Basic Componential Analysis of Space Heating Systems and Subsystems

6.7.2.1 Conceptualization of Heat Containment

Heat storage should be added between specific heat sources and distribution systems. Heat can be stored by sensed or absorption of heat, or through thermal decomposition reversible chemical reactions. However, phase change materials (PCM) store latent heat by using the fusion energy (for melting/solidification). Pit thermal preservation, which consists of excavation ditches filled with rocks and fluid, may be used in huge solar systems. The quantity of retained heat, thermal performance, and accessible energy all contribute to heat storage quality. Heat-producing energy is stored in temperature-controlled heat storage. This means that the temperatures in heat storage vary. The heating of water to a constant temperature in combined heat storage is carried out, which causes energy losses (Harvey 2012) (Figure 6.9).

Based on the duration of application, heat storage for space heating systems can be characterized as follows:

1. An armoured metal container of one or more heating systems is used for short-term heat storage. For excessive solar storage systems that require fallback air-conditioning, multiple heat converters are appropriate. An electrically refractory heater can be used as backup heating in smaller setups. They provide home hot water heating (DHW).
2. The midterm heat storage capacity has increased and is connected to space heating. Includes a DHW temperature conditioning system. Heat storage can indeed be linked directly or indirectly to a thermal source. Temperature stratification is essential because of poor winter light irradiation and high-water heating temperatures.

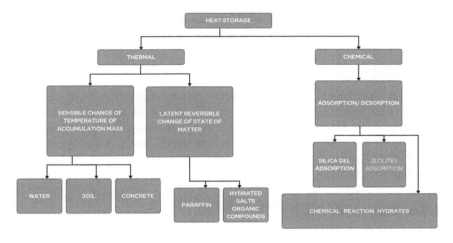

Figure 6.9 Heat storage types and bifurcations (Ostry 2013).

3. Long-term or periodic heat storage allows heat to be stored during the summer and released during the heating season the following year. Such storage is frequently used in solar-assisted district heating. Two technologies are commonly employed for large-scale systems: (i) a liquid variety consisting of enormous concrete barrier chambers containing water, and (ii) a pits type comprising of voids supplied with liquid and palm rocks.

6.7.3 Distribution Systems

A redistribution system is a collection of tubes or ducts that distribute thermal storage fluid between heat producers, energy storage containers and temperature transmitters in the space. It needs to be constructed in such a way that: (i) the cost of the distribution system is minimized; (ii) the expense of supplementary power generation for heat transfer and maintenance costs are minimized; (iii) heat losses from pipes or channels are minimized.

6.7.4 Hydronic Systems

A grid of pipes with steel, copper or plastic fittings is used in hydronic heating systems. These pipes are simple to install and have smooth, low-friction surfaces. Pumps, safety devices (expansion tanks, control valves, air filters) and control equipment are all part of the distribution system.

Water flow rate is affected by process variables. Flow rate can be controlled by manual, motor-driven or thermostatic valves. Flow rate decreases as pipe pressure decreases. The energy usage of the pumps will not vary. Lowering the rotor's rotations with a frequency-controlled electric motor improves efficiency. This necessitates the use of a device that reduces the frequency of electrical current to less than 50 Hz.

6.7.5 End Heat Exchangers

End heat exchangers are devices that transmit heat flux from the heat transfer fluid to the inside air. They are also known as heat emitters, which is another name for them. Examples of heat emitters include plate radiators, convectors, underfloor heating, ceilings or walls, along with dynamically actuated architectural structures. Heat emitters also include thermally actuated architectural structures.

6.7.6 Radiators

Radiators are built of metals (steel or aluminium) and release heat flux by convection and radiation. Radiators were formerly put beneath windows to avoid surface accumulation of water vapour within glass, to limit the flow of downward cold air below the window (due to cold glass and poorly sealed windows) and to increase the mean radiant temperature (Figure 6.10).

6.7.7 Active Beams

Rather than using fan coils, hydronic zone heating and cooling can be connected using active beam units, which operate as the termination units of a ventilation system devoted to developing fresh adequate ventilation in the room. A plenum gets controlled adequate ventilation from the primary AHU in an active beam unit. New air is forced via a series of nozzles at high velocity, forcing air from the room to ascend through a coiled heating element, heat up and combine with clean air. In this technique, convection completely transmits a thermal gradient into the room.

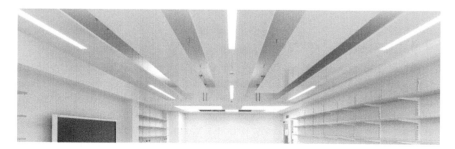

Figure 6.10 Ceiling radiant panels are generally used in nonresidential structures for heating and cooling purposes (SPC 2017).

6.8 THERMALLY ACTIVATED BUILDING STRUCTURES

Thermally activated building structures (TABS) are structures that employ pipes placed into the heart of the ceiling structure to maintain a nearly constant temperature of 20–24°C all year. Such systems enable cooling and large heat luminous cooling if the building's width is adequate (25–30 cm), a large heat volume material in building projects (e.g., concrete) is used, and floors on both sides are not encased by a ceiling building's thermoelectric insulating material to ensure appropriate open accrual thermal mass.

6.8.1 Space Heating System: Control

Since interior HVAC systems rarely operate at operation conditions, they must be managed correctly and adjusted to the current temperature variation of the structures. Controlling a space heating system entails altering one or even more characteristics, such as the temperatures, flow rate or tension of the heat exchange liquid in the supply system. Sensors and actuators comprise the control system. For successful control, a regulatory mechanism is also needed. It might be incorporated inside an actuator, a standalone IT device that connects several actuators and sensors.

6.8.2 Domestic Hot Water Heating (DHW) in nZEB

Domestic hot water is required for basic grooming and cleaning (DHW). In reduced and receptive structures, energy used for DHW water heating may exceed the energy used to power the structure. Domestic hot water technologies are included in the energy consumption evaluation of buildings as part of the Energy Performance of Building Directive systems. The wise usage of DHW has a substantial influence on both energy efficiency and environmental conservation. Reduced DHW consumption decreases the quantity of drinking water consumed, which is piped from freshwater sources or treated in water-cleaning facilities before even being supplied to structures. Reduced DHW use reduces the overall amount of wastewater required to be processed in centralized water filtration utilities (Aydinalp 2004). Unlike interior heating, the energy used for DHW preparation is continuous throughout the year. As a result, DHW solar heat pumps greatly enhance two nZEB indicators: lower primary energy demand for building operation and increased percentage of renewable energy sources. Solar Dehumidifying heating is currently mandated in some nations or towns, and such devices are supported financially in the bulk of European countries. Because the fluid is the heat transmission fluid in most situations, such systems can only be utilized in mild temperature zones or during certain times of the year to avoid the water from freezing. Closed-loop control solar heat pumps are widespread in continental climatic zones. Solar collectors and thermal mass are linked by a circulation

92 Thermal Energy Systems

tube in such systems, and pump and monitoring units are built to assure heat transfer fluid circulation. This is usually a blend of water and frost-protective fluid. The control unit monitors the thermostat in the solar collectors and heat storage but also activates the pump when there is a positive difference in temperature (Aydinalp 2004).

6.8.3 Space Setup Cooling Conceptualization for nZEB Structures

A cooling system, as well as warming, and lighting systems, are not required in a building because thermal environment specifications can be encountered through increased indoor thermal concentration mass, green-making buildings, effective colouration of glass windows as well as other translucent frameworks, or/and strenuous evening ventilation. However, however order to provide thermal comfort in any indoor or outdoor situation, a cooling system is required. In such cases, the energy consumed by a cooling system must be incorporated into the 's energy performance criteria (Guillén-Lambea 2016; Medved 2019).

6.8.4 Mechanical Cooling Systems and Concepts for nZEB

Mechanical cooling transfers heat from the cooler portions of the structure to the warmer outside. A refrigerator thermodynamics cycle is employed in line with the second law of thermodynamics. Install nZEB space cooling. Vapour compression chilling is widely used. COP is computed empirically in labs during continuous working circumstances and therefore is lower than theoretical. Because solar radiation intensity combined with domestic thermal mass causes the structure's cooling demand to fluctuate daily and periodically, chillers are usually only running at half capacity. The conditioning element load ratios of chillers vary. The COP of a chiller fluctuates during partial load. The fundamental concept is to swap the blower with a circulatory compressor that consumes less energy and transforms part of it into heat. Because the generation of energy has a greater environmental impact than the creation of heat (Guillén-Lambea 2016).

Straight vaporization (DX) or cold-water systems can be used to cool building mechanical spaces. Based on the kind of cold transfer fluid utilized within the building, chilled water systems employing compressors or thermally driven cooling systems are classed as all-air, all-water or mixed systems. While DX systems can be decentralized (cooling one room at a time) or central (cooling multiple rooms at once), and the cooling system is the cooler, transfer viscous liquid all throughout refrigeration unit, refrigerated liquid cooling systems are typically central, and water is usually allowed to cool to 5–7°C inside the cooler. Chilled water is subsequently transported through its supply system into the air handling units (Air handler) in all-air cooling either end-heat source in rooms across all coolers.

6.8.5 Direct Evaporation Based Air Conditioning Systems

When you notice the letter 'DX,' it signifies the fact that there is direct heat transfer between the interior air and the nitrogen in the evaporator. Most DX systems are split, which means that even an exchanger is located in the chilled area (room), while the remainder of the cooling elements is installed outside. It's the most frequent sort of DX system architecture. A single outside unit is connected to a number of refrigeration systems placed within the building in multi-split systems. This allows for adaptive management of the heat of the air quality in each particular room. Pipes connect the interior and external units, which both create the refrigerant loop, which circulates gas refrigeration (in the form of a fluid travelling forward and a gas travelling backward). The following are the advantages of DX systems: (i) effective legislation of required ambient temperature through various rooms; (ii) removal of air particles (dust atoms) by filter built in HVAC systems; (iii) portable refrigerant sequence piping that may be extensive (several metres) with a longer exposure (up to 50 inches); (iv) refrigerant or before aspects; (v) significant percentage of air conditioning units that are coupled (Medved 2019).

6.9 VENTILATION SYSTEMS AND MANAGEMENT IN nZEB

Fluids flow in nature when there is a pressure difference between regions: inside the instance of structures, it is between the interior and outside. Fans are used in regulated aeration to feed and withdraw air at the same time. To reduce ventilated thermal losses or gains, regulated ventilation is needed to recuperate the energy or coldness from extracting air to fresh air. Hybrid circulation is a regulated blend of both methods that is used to reduce energy demand for fan operation and building overheating.

6.9.1 Natural Ventilation

6.9.1.1 Working of Natural Ventilation Subsystems

Natural ventilation can indeed be powered by buoyancy or by wind. Neutrally buoyant ventilation occurs as a result of density differences between warm and cold air, and hence variances in hydrostatic air pressure. The pressure of the mass of the air column acting in a direction parallel to the observable plane causes hydrostatic air pressure. If the building's envelope is not tight or if ventilation apertures exist, changes in static pressurization cause a mass of air to flow into and out of the structure in the direction of higher to lower hydraulic air pressure. The pressure neutrality level is the point in a room when the exterior and inside hydrostatic air pressures are equal (Figures 6.11 and 6.12).

94 Thermal Energy Systems

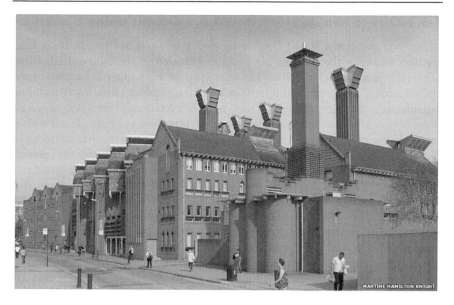

Figure 6.11 A real-life example of a naturally ventilated building.

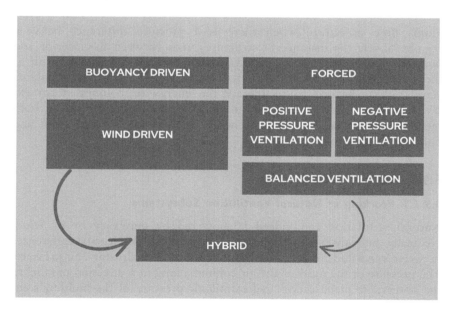

Figure 6.12 The schematic diagram depicts the various principles inculcated under building ventilation processes (Chartier 2009).

6.9.2 Mechanical Ventilation

For mechanical ventilation, fans produce pressure differences. Natural ventilation cannot be as precisely controlled as mechanical ventilation. The air temperature and the wind's speed have little effect on mechanical ventilation. Filters included in such air remove impurities, and silencers reduce outdoor noise transmission. Balanced ventilation using a wind exchanger between isolate as well as air intake ductwork can minimize ventilation heat losses by 90% or more. Humidified supply air improves heat recovery efficiency when a heat exchanger is made of vapour-permeable material. Devices that produce, vibration, especially unstable air velocity in ducts must be eliminated in mechanical ventilation systems. Due to moderate wind speed in transmission ductwork, mechanical ventilation units demand large systems and AHU space. Mechanical ventilation systems use gas and ought to be built to use as little as possible (Figure 6.13).

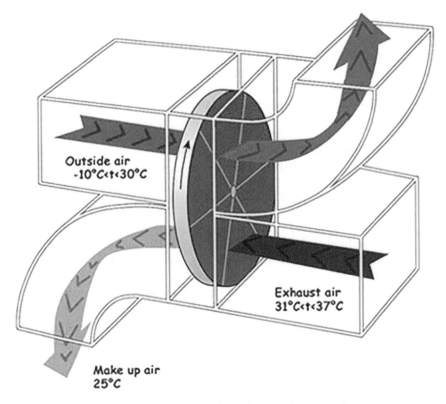

Figure 6.13 A rotary heat exchanger (mechanical ventilation system).

6.9.3 Mechanical Ventilation Systems

Mechanical ventilation systems remove stale air and give fresh air to spaces in a building or a home. Mechanical ventilation and energy recovery systems, for instance, remove and feed. Ventilators systems are divided into four categories:

- MVHR – Mechanical Ventilation Heat Recovery System
- C-MEV – Centralized mechanical extraction system, most known as mechanical extraction ventilation (MEV)
- D-MEV – Decentralized mechanical extraction system
- PIV – Positive input ventilation system

6.9.4 Heat Recovery Units (HRU) and Air Handling Units (AHU)

The heat efficacy of heat recovery is determined by the configuration of the HRU as well as the source and retrieve air flow rates. If the fluxes are equal, the greatest thermal effectiveness may be achieved. If HRU is made of a water vapour porous layer, the specific heat capacity of vapours is exchanged seen between extract with supply air in addition to perceptible heat. During the heating season, because the humidity of air pollutants is higher than that of the outside air, the specific heat capacity of liquid water is carried from the exhaust to supplying air, resulting in better heat recovery efficiency.

i. **Cross-Flow HRU:** Flows in counter-flow exchangers are perpendicular to one another. The pressure vessel exchanger is an example of cross-flow because one fluid flows within the tube side and the other water moves around the tubes on the shell side. When contrasted to the stream in the tubes, the flow surrounding those tubes seems to have a streamline of 90-degree angle. Cross-flow exchangers, like shell and tube exchangers, are commonly utilized in two-phase circuits. A condensing in a steam power plant is an example of an application in which steam travels through the tubes on the cylinder walls and changes form to liquid water. Coolant flows through the tubes, absorbing the heat from the steam.

ii. **Counter-Flow HRU:** In counter-flow heat exchangers, exhaust and supply airflow in opposing directions. Longer units have a higher pressure drop, requiring more fan energy. This heating system can theoretically be 100% efficient. To facilitate nighttime ventilation cooling, the AHU must include a supply airflow contained in the en route (all-around heat exchanger). Instead of aluminium, exchanger panels could be built of pore-structured polymer. This results in the formation of enthalpy heat exchangers. Water vapour molecules are also transferred between the extract and air distribution flows via osmosis. Because vapour atoms change from extraction to supply gas in the winters and vice versa in the summer, warmer air of water vapour improves the energy efficiency (equivalent to 120% as a result compared to standard HRU) and interior thermal comfort (Figure 6.14).

iii. **Flip-Flop HRU:** Regenerative heat exchangers are flip-flop heat exchangers. Unlike rotating wheels, highly porous steam accumulation common people operate as regenerating heat exchangers; supplied and removed air are regularly routed over one of the exchangers by a valve or valve. In the beginning half of the time, removed air moves through the first gasifier, while supplied air flows through the second regenerator in the wrong manner. During the second part of the cycle, the airflow direction is switched, with air flow cooling the first regenerator and extract air heating the second (Mardiana 2012; Gao 2015) (Figure 6.15).

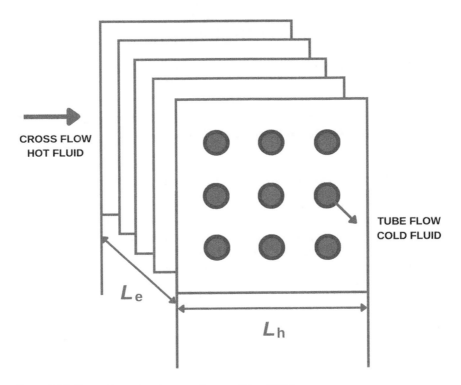

Figure 6.14 Cross-flow plate heat exchanger (Gao 2015).

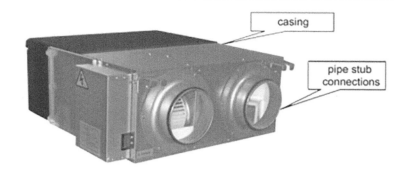

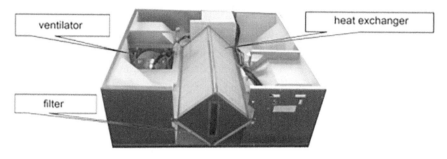

Figure 6.15 A compact HRU unit (Alnor Ventilation Systems, 2022).

6.10 TECHNOLOGICAL ASPECTS OF THE IMPROVEMENT OF VENTILATION ENERGY EFFICIENCY

6.10.1 Increment in Thermal Recovery-Based Prospects by the Utilization of a Ground Heat Exchanger

As or before (or pre-cool) supply air, a floor converter could be utilized. Horizontal conduit heating systems on these systems utilize atmospheric soil warmth, increasing the quantity of sustainable power in use in the construction phase. The air's temperature might well be boosted by 5–10°C. During the summer, GHX can pre-cool air flow in the daytime if the HRU is in a by-pass condition; at nighttime, a GHX by-pass operation is necessary to avoid air supply reheating leftovers.

6.10.2 Increasing Energy Efficiency of Buildings With Integration of Building Service Systems

Heat recovery may be accomplished if an air circulation system is established by incorporating a heat pump condenser into the exhaust channel.

Heat has been most likely used in smaller units for (pre)heating residential water, and in larger structures for space heating. Exhaust air ought not be reduced below 5°C to prevent moisture and freezing on the evaporator.

Heat can be relayed to other building products, like the DWH. In this scenario, DHW backup heating is required. A heat pump can also be put in a balancing mechanized ventilating subsystem, but only if the input air is thermally activated within the underground heat exchanger (Figure 6.16).

To summarize the chapter, the concept has been elaborated in terms of the development and operation based on thermal technological applications. Worldwide research takes us to a conclusion that the thermal applications for the development of nZEBs as a part of constructing sustainable structures for mankind are under the umbrella of applied thermal sciences and technology and holds great relevance and significance with respect to thermal comfort, technological development for heating, refrigeration and air conditioning systems. The trend moreover, goes

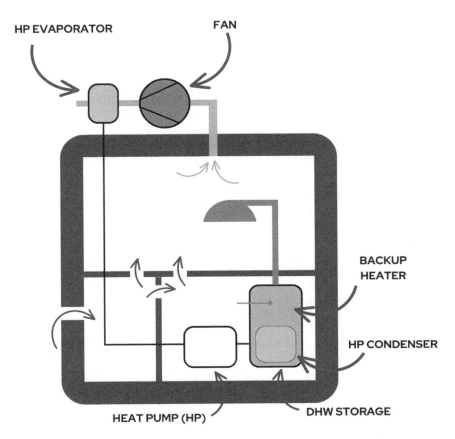

Figure 6.16 Mechanical ventilation of a heat recovery system (Medved 2019).

towards modernized technology for thermal comfort inculcation and energy development and processing for nZEBs. In recent years, there has been a rise in the amount of attention paid to the concept of zero-energy buildings. Numerous nations have already committed themselves to achieve ZEBs as their long-term energy goal for building construction. ZEBs have the intriguing ability to significantly reduce energy use while also increasing the overall percentage of renewable energy when compared to other techniques for reducing the amount of consumption in the building sector. These strategies include a variety of various approaches. However, in order to fulfil the requirements and not fall short of the expectations, there is a requirement for a ZEB-defining framework that is generally accepted and a reliable 'zero' calculation methodology. The upcoming decade will possibly witness a humongous increase in research, development, implementation and commercialization of nZEBs as a part of sustainable development for the technologically advanced future.

REFERENCES

Arteconi, A., & Polonara, F. (2018). Assessing the demand side management potential and the energy flexibility of heat pumps in buildings. *Energies, 11*(7), 1846.

Asdrubali, F., D'Alessandro, F., Baldinelli, G., & Bianchi, F. (2014). Evaluating in situ thermal transmittance of green buildings masonries—A case study. *Case Studies in Construction Materials, 1*, 53–59.

Aydinalp, M., Ugursal, V. I., & Fung, A. S. (2004). Modeling of the space and domestic hot-water heating energy-consumption in the residential sector using neural networks. *Applied Energy, 79*(2), 159–178.

Alnor Ventilation Systems. (2022). Heat recovery ventilation – what is it. https://www.ventilation-alnor.co.uk/index/support/alnor-knowledge-base/heat-recovery/what-is-heat-recovery.html Retrieved 18 September 2022.

Chartier, Y., & Pessoa-Silva, C. L. (2009). In: Atkinson, J., Chartier, Y., Pessoa-Silva, C. L., Jensen, P., Li, Y., and Seto, Wing-Hong (eds). *Natural Ventilation for Infection Control in Health-care Settings.*Geneva: World Health Organization. ISBN-13: 978-92-4-154785-7

Fabbri, K. (2015). Indoor thermal comfort perception. *A Questionnaire Approach Focusing on Children*. New York City, NY, USA: Springer.

Flanders, S. N. (1994). Heat flux transducers measure in-situ building thermal performance. *Journal of Thermal Insulation and Building Envelopes, 18*(1), 28–52.

Gao, G., Geer, J., and Sammakia, B. (2015). Review and analysis of cross flow heat exchanger transient modeling for flow rate and temperature variations. *ASME Journal of Thermal Science and Engineering Applications, 7*(4), 041017.10.1115/1.4031222

Gao, T., Geer, J., & Sammakia, B. (2015). Review and analysis of cross flow heat exchanger transient modeling for flow rate and temperature variations. *Journal of Thermal Science and Engineering Applications, 7*(4), 041017. 10.1115/1.4031222.

Guillén-Lambea, S., Rodríguez-Soria, B., & Marín, J. M. (2016). Review of European ventilation strategies to meet the cooling and heating demands of nearly zero energy buildings (nZEB)/Passivhaus. Comparison with the USA. *Renewable and Sustainable Energy Reviews*, 62, 561–574.

Harvey, L. D. (2012). *A Handbook on Low-Energy Buildings and District-Energy Systems: Fundamentals, Techniques and Examples*. London: Routledge. 10.4324/9781849770293. ISBN9781849770293

Kabele, K. et al. (2012). Heating and cooling. Educational Package, IDES-EDU Master and Post graduate education and training in multi-disciplinary teams implementing EPBD and Beyond.

Lenz, B., Schreiber, J., & Stark, T. (2012). *Sustainable Building Services: Principles-Systems-Concepts*. Basel Switzerland: Walter de Gruyter.

Mardiana-Idayu, A., & Riffat, S. B. (2012). Review on heat recovery technologies for building applications. *Renewable and Sustainable Energy Reviews*, 16(2), 1241–1255.

Medved, S. (2014). *Gradbena fizika II (Building physics II)*. Ljubljana: Faculty of Architecture, University of Ljubljana.

Medved, S., & Černe, B. (2002). A simplified method for calculating heat losses to the ground according to the EN ISO 13370 standard. *Energy and Buildings*, 34(5), 523–528.

Medved, S., Domjan, S., & Arkar, C. (2019). *Sustainable Technologies for Nearly Zero Energy Buildings*. Switzerland: Springer International Publishing. 10.1007/978-3-030-02822-0

Nussbaumer, T., Czasch, C., Klippel, N., Johansson, L., & Tullin, C. (2008). Particulate emissions from biomass combustion in IEA countries. *Survey on Measurements and Emission Factors, International Energy Agency (IEA) Bioenergy Task*, Swiss Federal Office of Energy (SFOE), Zürich. ISBN 3-908705-18-5, page no 1–40.

Oland, C. B. (2002). *Guide to Low-emission Boiler and Combustion Equipment Selection*. The Laboratory.

Op't Veld, P. (2004). *RESHYVENT cluster project on demand controlled hybrid ventilation in residential buildings with specific emphasis on the integration of renewables: Final report*. Rotterdam: Cauberg-Huygen.

Ostry, M., & Charvat, P. (2013). Materials for advanced heat storage in buildings. *Procedia Engineering*, 57, 837–843.

Sharma, N. K. (2020). Sustainable building material for green building construction, conservation and refurbishing. *International Journal of Advances in Science and Technology*, 29, 5343–5350.

Singh, R., & Shukla, A. (2014). A review on methods of flue gas cleaning from combustion of biomass. *Renewable and Sustainable Energy Reviews*, 29, 854–864.

SPC. (2017). Radiant heating – What's fact, what's fiction? | SPC. Retrieved 18 September 2022, from https://www.spc-hvac.co.uk/2017/11/14/radiant-heating-whats-fact-whats-fiction/

Stack Effect Ventilation (2022). Retrieved 18 September 2022, from https://santacruzarchitect.wordpress.com/2015/05/16/stack-effect-ventilation/

Vallabhy, S., Arun Kumar, M., Bharath, V., Dhakshina Moorthy, E., & Jain, H. K. (2019). Design of uPVC windows for lateral wind loads sandwich with hurricane bars for multistorey structures. *International Research Journal of Engineering and Technology*, 6(3), 4473–4477.

Werner, S. (2017). International review of district heating and cooling. *Energy, 137,* 617–631.

Yu, B., Kum, S. M., Lee, C. E., & Lee, S. (2013). Effects of exhaust gas recirculation on the thermal efficiency and combustion characteristics for premixed combustion system. *Energy, 49,* 375–383.

Chapter 7

Modelling and Simulation of Thermal Energy System for Design Optimization

Arijit Kundu, Ashwani Kumar, Nitesh Dutt,
Varun Pratap Singh, and Chandan Swaroop Meena

CONTENTS

7.1 Introduction to Basic Principles of Modelling and Simulation of Thermal Systems 103
7.2 Modelling 106
    7.2.1 Types of Models 108
    7.2.2 Analogue Models 108
    7.2.3 Mathematical Models 109
    7.2.4 Numerical Models 115
            7.2.4.1 Solution Methodology in Numerical Modelling 117
    7.2.5 Physical Models 131
7.3 Integration of Different Models 131
7.4 System Simulation 132
    7.4.1 Steady or Dynamic Simulation 133
    7.4.2 Continuous or Discrete Simulation 133
    7.4.3 Deterministic or Stochastic Simulation 134
7.5 Methodology 134
7.6 Methods for Numerical Simulation 135
7.7 Optimization of Thermal Systems 136
References 138

7.1 INTRODUCTION TO BASIC PRINCIPLES OF MODELLING AND SIMULATION OF THERMAL SYSTEMS

A thermal system is described as a consortium of several thermal, fluid and power-generating elements with associated properties. Modelling and simulation of such a system are similar to observing a contrived system that mimics the functionality of an actual system. To ascertain the predicted operating circumstances and performance at design as well as off-load conditions at all operational points of the system, system simulations are carried out. By doing so, it is possible to establish the operating variables' limiting

DOI: 10.1201/9781003395768-7

103

circumstances. This can be applied to an existing design to investigate the possibility of performance-improving alterations, while it is typically used in the design stage to examine alternatives and/or an improved design. A system simulation demands knowledge of all mass and energy balance equations, constitutive relations and equations for the thermodynamic properties of the working substances, and the performance characteristics of all components. A set of simultaneous equations is made up of the thermodynamic properties, equations for how well each component performs and equations for the conservation of mass and energy. It is typical to conduct a system simulation analysis during the design phase to create a better design and investigate the possibilities of altering an existing design for greater performance and economics at the limiting conditions of the operating variables.

In present practice, words like "model" and "simulator" are frequently used in vague ways, but they have very clear definitions in the framework we will explain. Therefore, it's critical to comprehend what the definitions cover and exclude. The framework's fundamental components are the source system, model, simulator and experimental frame, as shown in Figure 7.1. Modelling and simulation relationships are the fundamental connections between entities [1].

The real or artificial data from the source, that we are interested in modelling, are collected and gathered in a parameter database and observed or experimented with the conditions under which the system is established. In the form of time-indexed variable trajectories, it is seen as a source of observable data. An experimental frame is a description of the circumstances

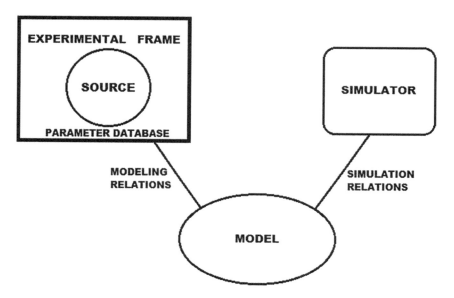

Figure 7.1 The basic entities in modelling and simulation and their relationships.

under which a system is observed or subjected to an experiment. The operational definition of the goals that drive a modelling and simulation project is as a result an experimental frame. Through modelling relations, the model describes the directions for producing such data, and simulator, a computational device, generates the final behaviour of the model under these behaviour parameters of the system. There are still many different modelling and simulation terminologies, as well as concepts with many different interpretations, even after several decades of development.

The role that a model plays in system design, management, or control is related to modelling objectives. The objective statement directs model construction toward specific problems. Such a statement must be created as early in the development process as is practical. Project managers can keep tabs on the team's actions thanks to a clearly defined set of objectives. It is possible to create appropriate experimental frames if the objectives are established. You should keep in mind that these frames translate the objectives into more accurate experimentation settings for the underlying system or its models. It is predicted that a model will hold true for the system in each of these frames. After stating our goals, it is likely that there is an optimum level of resolution to address the issues mentioned. The more the resolution presumably required answering the issue, the more difficult the question is. As a result, the objectives and the experimental frame counterparts have a crucial role in the decision of suitable levels of abstraction.

An essential phase in the design and optimization of the system is the modelling of the physical system, which is produced from conceptual design and from the formulation of the design challenge. In order to simplify the given problem and make it possible to investigate the characteristics and behaviour of the system for various conditions, it is necessary to concentrate on the dominant aspects of the system while ignoring relatively small effects because most practical thermal systems are fairly complex. The processes that control the system are idealized and approximated in order to make the study simpler. Modelling of thermal systems also heavily relies on fundamental conservation laws and material characteristics. Because of the complexity of fluid flow and the mechanisms that control heat and mass transport in thermal systems, the study of these systems is frequently challenging. Because of this, it is necessary to estimate, simplify and idealize common thermal systems in order to study them and get the information essential for design. The characteristics listed below are some of those that thermal systems and processes frequently confront [2]:

1. Time-dependent
2. Multidimensional
3. Complex geometries
4. Complicated boundary conditions
5. Coupled transport phenomena

6. Turbulent flow
7. Change in phase
8. Energy losses and irreversibility
9. Variable material properties
10. Influence of ambient conditions
11. Ambient circumstances' impact
12. Variety of energy sources

The governing equations are often a set of partial differential equations, with nonlinearity emerging due to the convection of momentum in the flow, changeable characteristics and radiative transport. This is because typical systems are time-dependent and multidimensional in nature. But in order to make these equations easier, approximations and idealizations are utilized, leading to algebraic and ordinary differential equations for many real-world scenarios and relatively simpler partial differential equations for others.

7.2 MODELLING

One of the most important aspects of the design and optimization of thermal systems is modelling. In order to make a problem acceptable to a solution, practical processes and systems must be simplified through idealizations and approximations. Modelling is the process of making a problem simpler so that it can be expressed as an equation system for analysis or a physical configuration for investigation. Models are used to get pertinent quantitative inputs for designing and improving processes, parts and systems. Understanding and making predictions about the behaviour and properties of thermal systems need modelling. A model is acquired and then put through a number of operating scenarios and design iterations. The outputs from the model characterize the behaviour of the given system if it accurately represents the real system under consideration. This information is utilized both in the design process and in the evaluation of a specific design to see if it satisfies the requirements and limitations that have been established. By projecting the performance of each design, modelling also aids in acquiring and evaluating alternatives, and ultimately results in the best design. As a result, the design and optimization processes are directly related to the modelling effort, and the validity and accuracy of the model used has a significant impact on the outcome of the final design.

Because of the generally complex nature of the transport brought on by variations with space and time, nonlinear mechanisms, complicated boundary conditions, coupled transport processes, complicated geometries, and variable material properties, modelling is particularly important in thermal systems and processes. As a result, sets of time-dependent, multidimensional, nonlinear partial differential equations with intricate domains and boundary conditions frequently govern thermal systems. Finding a solution to the entire time-dependent, three-dimensional problem is typically a very laborious procedure. The huge number of variables involved also makes it typically difficult to comprehend the results and apply them to the design process. Even while experiments are conducted to get the necessary input data for design, the cost associated with each experiment makes it essential to create a model to direct the experimentation and to concentrate on the key variables. As a result, it's essential to ignore relatively insignificant details, aggregate the effects of many problem factors, use idealizations to make the analysis simpler and minimize the number of parameters that control the process or system.

The model could be descriptive or predictive. Models that are used to describe and explain diverse physical events are something we are all quite familiar with. To demonstrate how a mechanical system works, such as a water pump, heat exchanger, robot or internal combustion engine, a working model is frequently utilized. The model frequently has a cutaway part that reveals the internal mechanisms or is made of clear plastic. These models, known as descriptive models, are commonly applied to clarify fundamental mechanics and guiding ideas.

Because they can be used to forecast a system's performance, predictive models are especially relevant to our current topic of engineering design. Because it enables us to determine the dependence of the cooling curve on physical variables like the initial temperature of the sphere, water temperature and material properties, the equation governing the cooling of a hot metal sphere immersed in an extensive cold-water environment represents a predictive model. In Figure 7.2, a couple of these models are sketched. To understand the fundamental concepts and derive the governing

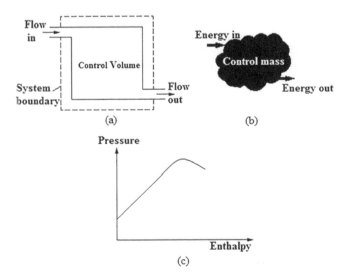

Figure 7.2 A few models commonly used: (a) control volume, (b) control mass, and (c) commonly graphical representation [3].

equations, engineering analysis frequently uses models like the control mass and control volume formulations in thermodynamics, the representation of a projectile as a point to study its trajectory, and enclosure models for radiation heat transfer.

The validation of the model created for a particular system is another crucial factor since it establishes if the model is an accurate depiction of the real physical system and how accurate the model's predictions can be expected to be. The physical behaviour of the model, its application to current systems and processes, and comparisons with experimental or numerical data are frequently the foundations for validation. Because modelling and design are interconnected, the model can be improved using the results of system simulation and design. Models are initially created for certain actions and elements, and then these separate models are coupled to create the model for the complete system. The governing equations, connecting equations obtained from experimental data, curve-fit findings from data on material qualities, characteristics of pertinent components, financial trends, environmental aspects, and other design-relevant concerns typically make up this final model.

7.2.1 Types of Models

A thermal system can be modelled using a variety of different sorts of models. Each model has unique qualities and is best suited for various situations and purposes. In the part before that, it was discussed how models might be classified as descriptive or predictive. Predictive models that can be used to forecast a system's behaviour under various operating situations and design parameters are of particular interest to us. Defined as the process of creating such models, modelling here shall only be used to describe predictive models. The design and optimization of thermal systems are particularly interested in four different categories of predictive models, such as [3]:

1. Analogue models
2. Mathematical models
3. Numerical models
4. Physical models

7.2.2 Analogue Models

Analogue models depict the physiological process by employing components that are, to some extent, comparable to those in the actual process. The usage of analogue models, which are based on the analogy or likeness between various physical processes, enables one to apply the solution and output from one known problem to another that has not yet been addressed. Secondary characteristics like energy utilization, which is typically similar between analogous elements and the real components they represent, are better represented by analogue models. In the fields of thermal energy systems, analogue models are frequently used [4]. Conduction heat transfer through a composite wall serves as an illustration of an analogue model. It can be studied in terms of an analogous electric circuit, where the thermal resistance is represented by the electrical resistance and the heat flux is represented by the electric current, as shown in Figure 7.3. The analogue model must adhere to the same physical laws even when it obviously differs physically from the system under examination.

Analogue models, however, have only a limited application in engineering design. They are valuable in understanding physical phenomena and in expressing information or material movement. This is primarily due to the fact that the analogue models itself must be solved and may have similar difficulties to the original issue. As a result, it is often preferable to create a proper mathematical model for the thermal system rather than adding an analogue model to it.

7.2.3 Mathematical Models

A mathematical model typically explains a system using a collection of variables explaining the characteristics and a set of equations that construct relationships between the variables. These models are the most crucial ones for designing thermal systems because they offer a great deal of flexibility in acquiring the quantitative data required as design inputs. Numerical modelling and simulation are built on mathematical models, allowing for the investigation of the system without the need to create a physical prototype. The mathematical model's simplifications and approximations also

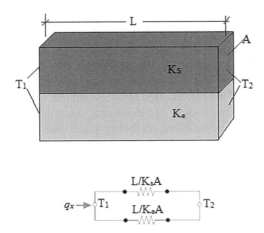

Figure 7.3 Analogue models of conduction heat transfer in a composite wall.

110 Thermal Energy Systems

reveal the key variables in a situation. If necessary, this aids in the development of effective experimental models. For instance, the total heat transfer rate's dependency on the input temperatures of the two fluids and the system's dimensions would be revealed by solving the equations for a heat exchanger. Similar to this, the solution of the associated governing equations yields the dependence of the solidification time in casting on the beginning temperature and cooling conditions. These findings serve as the foundation for design and optimization. Numerous purposes can be served by mathematical modelling. The level of system knowledge and the quality of the modelling determine how successfully any given target is accomplished. The model could have a physical underpinning or be based on curve fitting of numerical or experimental data. These two methods produce two different kinds of models, which are frequently referred to as theoretical and empirical, respectively. Convective transport from heated bodies of various shapes is represented by heat transfer correlations, which are empirical models that are extensively used in the development of thermal systems. Empirical models are produced through curve-fitting data to produce mathematical representations of experimental or numerical outcomes.

In mathematical modelling, compromise plays a significant role. In the real world, the bulk of interacting systems is far too complex to model in their entirety. Therefore, identifying the most crucial components of the system becomes the first level of compromise. The remainder will not be incorporated into the model; just these will. The second degree of compromise deals with how much mathematical manipulation is actually beneficial. Although general conclusions can be proved using mathematics, these conclusions are very dependent on the type of equations that are utilized.

We must be certain of our goals before starting a modelling assignment. These influence the project's future course in two different ways. First of all, the use of the model will determine the level of detail that is included. Separating the system to be modelled from its environment is the second step. If the environment influences the behaviour of the system but the system has no impact on the environment, then this division is valid. After deciding which system will be modelled, we must build the model's fundamental structure. This reflects our opinions on the way the system works. These ideas can be expressed as underlying assumptions. Future system analysis takes these hypotheses as being true, but the accuracy of the analysis' findings depends on the assumptions.

Whether the system may be believed to be in a steady state, involving no variations with time, or whether the time-dependent changes must be taken into account, is one of the most crucial factors in modelling. Determining whether these impacts may be disregarded is crucial since adding time makes the task more complex by introducing another independent variable. Although the majority of thermal processes rely on time, they can be roughly characterized as constant under some practical conditions. Even while every occurrence begins as a transient issue, it eventually moves toward a steady-state situation. For instance, it is obvious that the solar heat flux incident on a house's wall changes with time. Nevertheless, it can be roughly described as steady for a few brief times.

Two main characteristic time scales need to be considered. The first, t_r, refers to the response time of the material or body under consideration, and the second, t_c, refers to the characteristic time of variation of the ambient or operating conditions. Therefore, t_c indicates the time over which the conditions change. For instance, it would be zero for a step change and the time period t_p for a periodic process, where $t_c = 1/f$, with f being the frequency. The response time t_r for a uniform-temperature (lumped) body subjected to a step change in ambient temperature for convective cooling or heating is given by the expression [3]:

$$t_r = \frac{\rho C V}{h A}$$

where ρ is the density, C is the specific heat, V is the volume of the body, A is its surface area and h is the convective heat transfer coefficient. These two-time scales can be used to obtain a number of significant situations, including the following:

1. **t_c is very large, i.e., $t_c \to \infty$:** In this scenario, it is possible to assume that the conditions won't change over time and treat the system as steady state. Transient impacts are significant at the beginning of the process when the variables change dramatically over a brief period of time. But steady-state conditions are reached as time goes on.
2. **$t_c \ll t_r$:** In this instance, the operating conditions change far more quickly than the material's response. After that, the material becomes unable to adapt to changes in the operational factors. When the operating circumstances are considered as their mean values, the system may then be roughly described as steady.
3. **$t_c \gg t_r$:** This is a situation in which a substance or body responds quickly yet operational or boundary conditions vary gradually. The surroundings can then be assumed to have remained constant for a portion of the associated response time. As a result, the part may be described in such circumstances as quasi-stationary, with the steady problem being solved at various points.
4. **Periodic:** The characteristics of the thermal system's activity can be described as a periodic process that repeats itself over a predetermined amount of time t_p. These circumstances could be portrayed as:

$$\int_0^{t_p} \varphi(t)\,dt = 0; \quad T_t = T_{t+t_p};$$

where $\varphi(t)$ is the physical or thermal property of the body as a function of time. A periodic process can be modelled using one of the afore-mentioned requirements. The governing equations still contain the time-dependent elements, and the issue is worked through until the system's cyclic behaviour is attained. Because the process is periodic, analytical solutions can frequently be found, especially if the periodic process can be roughly represented by a sinusoidal variation [5].

5. **Transient:** The system must be represented as a general time-dependent issue with the transient factors incorporated if none of the aforementioned approximations apply. Since this is the situation that presents the greatest challenge in terms of time dependence, attempts should be taken to make it simpler before turning to complete transient or dynamic modelling.

The next factor is figuring out how many spatial dimensions are required to model a particular system. All realistic systems are three-dimensional, although to greatly simplify modelling, they are frequently approximated as two- or one-dimensional. The geometry of the system under study and the boundary conditions serve as the foundation for this significant simplification. The dimensionless variables can be used to generalize and simplify the equations, make them irrespective of a particular system of units. Scale analysis, which is based on taking into account the scales of the many quantities involved, yields conclusions that are similar [6]. Even though a given system is essentially a three-dimensional problem, it can be simplified in modelling by being represented as one-dimensional, two-dimensional, or axisymmetric. Due to the increased difficulty in solving the governing equations, three-dimensional modelling is typically avoided until absolutely necessary. Additionally, three-dimensional model findings are difficult to comprehend and frequently need for specialized methods.

Mathematical models have been created by physicists to describe a variety of systems. The application of mathematical equations is frequently straightforward since the systems may frequently be stated accurately. The choice of mathematical equations to describe the system must be made when the framework of a model has been established. It is important to choose these equations carefully since they could have unanticipated consequences on how the model behaves. It can be maintained to produce a very straightforward model known as the lumped mass approximation. The temperature, species concentration, or any other transport variable is expected to be uniform within the domain of interest in this model, which is widely employed and is thus a significant situation. As a result, the variable is grouped and the region's spatial variation is not taken into account. Instead of differential equations, algebraic equations are produced for steady-state situations. The conditions in the various components of the majority of thermodynamic systems, including air conditioning and refrigeration equipment, internal combustion engines, power plants, etc., are assumed to be uniform and so grouped when they are examined [7]. The lumped mass approximation is commonly employed in modelling because of the significant simplification it produces and the fact that it frequently accurately depicts the process.

Practical systems and processes typically have challenging, irregular, and time-varying boundary conditions. To achieve a large degree of simplification, the borders should be approximated as smooth, with simpler geometry and homogeneous circumstances, without significantly sacrificing accuracy or generality. To greatly simplify the model, conditions that change throughout the boundaries or over time are frequently treated as uniform or constant. These boundary condition simplifications not only make the model less complex but also make it simpler to comprehend and extrapolate the model's findings. By ignoring relatively minor effects, significant simplifications in the mathematical modelling of thermal systems can be achieved. Considerations that are of minimal importance are eliminated using estimates of the pertinent quantities. Idealizations are frequently used to make problems simpler and to arrive at the best possible answer. The performance of actual systems can then be evaluated in terms of this ideal behaviour and expressed in terms of effectiveness, efficiency, or coefficient of performance. It is crucial to use precise material property data when mathematically modelling any thermal system or process. Physical factors like temperature, pressure and species concentration typically affect the characteristics. The governing equations for thermal systems and processes are derived using the laws of conservation of mass, momentum and energy as a starting point. Using the numerous factors mentioned in the previous discussion, the equations are made simpler. Equations may be algebraic, differential or integral as a result. Because they apply locally and enable the identification of variations in time and space, differential methods are the conservation formulation that is used the most commonly. There are a few hypothetical scenarios where only one independent variable is taken into account that leads to ordinary differential equations like transient models. For distributed models, partial differential equations are obtained. In very few instances, analytical methods can be used to solve the partial differential equations that govern the majority of real-world thermal systems; instead, numerical methods are typically required. The methods for partial differential equations that are most frequently used are the finite-difference and finite-element approaches. Analytical solutions to common differential equations are frequently possible, especially when the equation is linear. Further simplification can occasionally be obtained by examining the various terms in the equations to see if any of them are negligible after the governing equations and the pertinent boundary conditions have been formed using the various approximations and idealizations. The governing equations are typically nondimensionalized, and the governing parameters are assessed.

In mathematical modelling, it is most usual to approximate the boundary conditions of practical processes that entail complex, non-uniform, and time-varying boundary conditions as smooth, with simpler geometry, and with uniform conditions, without significantly sacrificing accuracy or generality. To greatly simplify the model, conditions that change throughout the boundaries or over time are frequently treated as uniform or constant. All such approximations must, however, adhere to the same fundamental conservation laws that apply to the given profile. These boundary condition simplifications not only make the model less complex but also make it simpler to comprehend and extrapolate the model's findings. By ignoring relatively minor effects, significant simplifications in the mathematical modelling of thermal systems can be achieved. Considerations that are of minimal importance are eliminated using estimates of the pertinent quantities. To quantify the range of variation of the pertinent parameters, such estimations frequently rely on data from existing, comparable processes and systems. Idealizations are frequently used to simplify problems, concentrate on important factors, avoid features that are frequently difficult to quantify and produce solutions that exhibit the best performance [8]. The performance of actual systems can then be evaluated in terms of this ideal behaviour and expressed in terms of effectiveness, efficiency or coefficient of performance.

In very few instances, analytical methods can be used to solve the partial differential equations that govern the majority of real-world thermal systems; instead, numerical methods are typically required. The methods for partial differential equations that are most frequently used are the finite-difference and finite-element approaches. Analytical solutions to common differential equations are frequently possible, especially when the equation is linear. The integral formulation, from which the finite-element and finite-volume approaches are derived, is based on an integral expression of the conservation principles and can be applied to the entire domain or a tiny finite region [9]. Further simplification can occasionally be obtained by examining the various terms in the equations to see if any of them are negligible after the governing equations and the pertinent boundary conditions have been formed using the various approximations and idealizations. The general basis for this is a non-dimensionalization of the governing equations and an evaluation of the governing parameters.

The mathematical modelling of a thermal system often entails modelling each of the system's constituent parts and subsystems, and then coupling all of these models together to produce the system's final, integrated model. Combinations of algebraic, differential, and integral equations may make up the governing equations. One component might be represented mathematically as a lumped mass, another as a one-dimensional transient, and yet another as a three-dimensional object. Using physical knowledge and estimations of the various transport

Modelling and Simulation of Thermal Energy System   115

processes, the overall mathematical modelling process may be applied to each of these components. In order to solve the mathematical equations and learn more about the behaviour and characteristics of the thermal system, we must first obtain the final model of the system. This approach involves simulating the system under a number of different operating and design situations. To better depict the physical system, we must evaluate the mathematical model and make any necessary improvements before moving on to simulation. The three most frequently used validation techniques are determining whether the model's activities is physically plausible, comparing the results to those of simpler systems, or comparing the results to the data from full-scale systems.

7.2.4 Numerical Models

Mathematical equations that describe the behaviour of the thermal system are derived as a result of mathematical and physical modelling, curve fitting, and other techniques. Common connections between these equations include material qualities, boundary conditions, material and energy flow, and interactions between the system's many parts. To ascertain the performance and characteristics of the system for a wide variety of design factors and operating conditions, we are interested in finding solutions to this coupled set of equations. Analytical solutions are rarely feasible due to the coupled nature of these equations, and since nonlinear algebraic and differential equations, including both ordinary and partial differential equations, arise in typical thermal systems, we must turn to numerical techniques in order to achieve the desired results. A computational or numerical representation of the thermal system on a computer is referred to as a numerical model, and it can be used to approximate the behaviour and features of the system. With numerically enforced boundary and beginning conditions, pertinent property data, component characteristics, and other inputs required for modelling the entire system, it consists of a numerical scheme or technique that would produce a solution to the governing mathematical equations. The numerical model is composed of the numerical algorithm and its computer implementation. The model is subjected to modifications in the design parameters and operational circumstances once it has been shown that it is a reliable and accurate representation of the system. Simulation is the process of study of a system's conduct using a model rather than creating a prototype.

Numerical modelling's primary goal is to create a computational code that can be implemented on a digital computer and gives a physically correct representation of the real system while also enabling the system's performance to be predicted under various circumstances. An exclusive correlation between the physical thermal system and the numerical model is established in order to obtain the necessary knowledge about system characteristics. This allows the numerical model to be run under various conditions. The first step in numerical modelling is the resolution of specific equations. The solution process for the collection of equations that regulate a particular portion or system is then produced by combining the numerical schemes for the various equations.

The numerical approach for solving each governing equation must be established after identifying the kind of each equation to be solved in accordance with the applied system. Then, using a familiar language of choice, a numerical code has been developed to answer each equation. To model a system element or subsystem, one must then run this code and combine results for numerous linked equations. To simulate the complete system, numerical models of each system component have been compiled. Finally, we must validate the accuracy of the complete system model. As a result, a methodical approach is employed to create an adequate numerical model for the entire thermal system. But experimental results and simplified analysis are often used to avoid solving some thermal system models because of the complexity of the resulting problem.

In numerical modelling, the degree to which the model captures the behaviour of the actual system is the most crucial factor. This factor takes into account both the model's predictions about the system's physical activities and the veracity of those predictions. Making sure the model is a physically sound and accurate depiction of the system is crucial. As a result, the produced findings ought to be independent of the numerical scheme and its application. The results must not be considerably impacted by the values used if arbitrary numerical parameters, such as grid size, time step, convergence criterion, and initial state points, are introduced in order to get a solution. Verification of the numerical system is another name for this. The physical features of the outputs, which must reflect patterns anticipated for the actual system, also serve as a critical check on the model's validity. When possible, comparisons of the results with experimental data are used to further validate the model and gauge the precision of the model's outputs. Truncation errors and round-off mistakes are examples of numerical errors. Truncation errors are brought on by the numerical scheme's approximations, such as those that swap out differential changes for finite ones, and round-off errors are caused by the computer system's limited precision. The equations, the beginning point, the convergence criterion, and the numerical scheme all play a role in whether or not an iterative scheme, which starts with an initial, guessed value and moves towards a solution, converges. As computing moves forward, the unpredictability of the scheme implies an infinite development in numerical errors. Determined and avoided are the situations where the plan becomes unstable or diverges. A convergent, stable and accurate numerical solution to a given mathematical problem is required in order to get relevant results from the numerical model [8].

Numerous computer programs are freely accessible and can easily be modified to work with the computer system and the overall numerical

Modelling and Simulation of Thermal Energy System   117

model. These programs include techniques for solving sets of linear algebraic equations, ordinary differential equations, and curve fitting and numerical integration. To make essential alterations and to adapt the inputs and outputs to the problem at hand, it is vital to comprehend the algorithm used by the programs that is currently available. The program's restrictions and the anticipated accuracy of the numerical findings should both be understood. It goes without saying that using these programs will greatly simplify computer programming. The use of such software is extremely common in engineering practice because the goal is frequently to get the desired results as soon as possible. These programs are often thoroughly tested and can be utilized to construct the numerical model. In industrial applications, general-purpose codes like Fidap, Fluent and Phoenics are frequently employed. MATLAB is a particularly well-liked piece of software that is frequently employed to resolve various mathematical problems in thermal system models.

7.2.4.1 Solution Methodology in Numerical Modelling

When modelling thermal systems mathematically, multiple types of governing equations must be solved using numerical methods for differentiation, integration and curve fitting. The set of linear and nonlinear algebraic equations, the ordinary and partial differential equations, and the integral equations are the various forms governing equations that are most frequently utilized.

It is crucial to solve several simultaneous linear algebraic equations. Numerous applications, including those relating to data analysis, conduction heat transfer, chemical reactions, fluid flow circuits and fluid flow circuits are frequently governed by linear systems. The general form of a set of n linear equations can be written as

$$
\begin{aligned}
\alpha_{11} x_1 + \alpha_{12} x_2 \ldots &\quad + \alpha_{1n} x_n = \beta_1 \\
\alpha_{21} x_1 + \alpha_{22} x_2 \ldots &\quad + \alpha_{2n} x_n = \beta_2 \\
\vdots \quad \vdots \quad \vdots \quad \vdots & \\
\vdots \quad \vdots \quad \vdots \quad \vdots & \\
\alpha_{n1} x_1 + \alpha_{n2} x_2 \ldots &\quad + \alpha_{nn} x_n = \beta_n
\end{aligned}
\tag{7.1}
$$

where α represents the coefficients, the x represents the n unknown variables, and the β represent the constants of the equations. The system may also be written more concisely in matrix notation as

$$
(a)(\mathbf{X}) = (b)
\tag{7.2}
$$

where (a) is an $n \times n$ square matrix of the coefficients where elements represented by α_{ij}, $(\mathbf{X})$ is a column matrix of the n unknowns with elements

118   Thermal Energy Systems

x_i, and (b) is a column matrix of the constants with elements β_i that appear on the right-hand side of the n equations.

With the exception of the round-off error, direct methods solve the equations precisely in a limited number of operations. Since several techniques rely on matrix inversion, the equation's solution can be expressed as

$$(\mathbf{X}) = (a^{-1})(b) \tag{7.3}$$

where (a^{-1}) is the inverse of matrix (a). This inverse cannot exist if the determinant of (a) is zero. When the determinant is zero, the equations are said to be homogeneous and the matrix (a) is said to be singular. Nontrivial solutions can only be found if the column vector (b) is likewise zero.

A direct method used to solve many engineering problems is Gaussian elimination. By creating a methodical plan or procedure to get rid of unknowns and back replace, the elimination process is codified and applied to vast sets of equations. The process turns the matrix (a) into an upper triangular matrix, leaving the bottom row empty. The equation corresponding to this bottom row is then a linear equation with just one easily determinable unknown. Back-substitution, which involves considering each row in turn while moving from the bottom to the top row, is used to find the remaining unknowns, resulting in each row having just one unknown. The column vector (b) is added to the end of the matrix to create an augmented matrix (a). The pivot row, also known as the element in the first column, and the pivot element, respectively, are both eliminated in the first step for all the rows below the first row. The second row becomes the pivot row in the second step, and the element in the second column becomes the pivot element. Once more, all of the components for rows below the second row are removed from the second column. Up until a higher triangular matrix takes the place of matrix (a), the operation is repeated.

Note that a division by zero may occur if the pivot element is zero. Gauss elimination does not check for this. Partial pivoting, which alternates the rows below and including the pivot row at each step to utilize the one with the largest pivot element as the pivot row, is frequently employed to prevent these issues [3]. The system is modified once eliminating x_1 as

$$\begin{bmatrix} \alpha_{11} & \cdots & \alpha_{1n} \\ \vdots & \ddots & \vdots \\ 0 & \cdots & \alpha'_{nn} \end{bmatrix} \begin{pmatrix} x_1 \\ x_2 \\ \vdots \\ \vdots \\ x_n \end{pmatrix} = \begin{bmatrix} \beta_1 \\ \beta'_2 \\ \vdots \\ \vdots \\ \beta'_n \end{bmatrix} \tag{7.4}$$

Then eliminating x_2 from 3rd to nth equations as

$$\begin{bmatrix} \alpha_{11} & \alpha_{12} & \cdots & \alpha_{1n} \\ 0 & \alpha'_{22} & \cdots & \alpha'_{2n} \\ 0 & 0 & \vdots & \alpha''_{3n} \\ \vdots & \vdots & \vdots & \vdots \\ 0 & 0 & \cdots & \alpha''_{nn} \end{bmatrix} \begin{pmatrix} x_1 \\ x_2 \\ x_3 \\ \vdots \\ x_n \end{pmatrix} = \begin{bmatrix} \beta_1 \\ \beta'_2 \\ \beta''_3 \\ \vdots \\ \beta'_n \end{bmatrix} \tag{7.5}$$

Repeating steps upto $(n - 1)$ times to yield upper triangular system

$$\begin{bmatrix} \alpha_{11} & \alpha_{12} & \cdots & \alpha_{1n} \\ 0 & \alpha'_{22} & \cdots & \alpha'_{2n} \\ 0 & 0 & \vdots & \alpha''_{3n} \\ \vdots & \vdots & \vdots & \vdots \\ 0 & 0 & 0 & \alpha_{nn}^{(n-1)} \end{bmatrix} \begin{pmatrix} x_1 \\ x_2 \\ x_3 \\ \vdots \\ x_n \end{pmatrix} = \begin{bmatrix} \beta_1 \\ \beta'_2 \\ \beta''_3 \\ \vdots \\ \beta_n^{(n-1)} \end{bmatrix} \tag{7.6}$$

Once the reduced matrix is obtained, the unknown x_n is obtained from

$$x_n = \frac{\beta_n^{(n-1)}}{\alpha_{nn}^{(n-1)}} \tag{7.7}$$

Once x_{n-1} has been calculated for each row above the bottom one, the remaining unknowns are then recovered through back-substitution. Thus,

$$x_i = \frac{\beta_i^{(i-1)} - \sum_{j=i+1}^n \alpha_{ij}^{(i-1)} x_j}{\alpha_{ii}^{(i-1)}} \tag{7.8}$$

where $i = n - 1, \ n - 2, \ .,1$.

There are other additional direct strategies for resolving collections of linear equations. The coefficient matrix (a) is converted to an identity matrix $(\mathfrak{I})$ using the Gauss-Jordan elimination method, which is a variant of the Gaussian elimination approach, with the only non-zero entries being unity along the diagonal. Since $(\mathfrak{I})$ $(\mathbf{X}) = (b')$ results in $(\mathbf{X}) = (b')$, the unknown vector $(\mathbf{X})$ is thus simply given by the modified constant vector (b'). Back-substitution is not necessary. The unknown is removed from both the equations above and below the pivot equation at each stage of the elimination process, and the pivot equation is then normalized by dividing it by the pivot element to produce an identity matrix at the conclusion of the procedure. It is possible to use a number of effective techniques based on matrix inversion or matrix decomposition into upper (u) and lower (ℓ)

120   Thermal Energy Systems

triangular matrices. This process is known as LU decomposition, and it can be accomplished using a variety of numerical techniques, including Crout's and Cholesky's approaches [3].

Equation (7.1) can be used to derive by the iterative strategy solving for the unknowns, starting with x_1 and yielding x_2, x_3, ..., x_i, ..., x_n in turn. For a large number of equations, elimination techniques like Gaussian elimination are susceptible to significant round-off errors. Iterative methods, such as the Gauss-Seidel method, give the user control of the round-off error, so very popular. Each equation is rewritten for the associated unknown if the diagonal elements are non-zero. For example, the first equation is rewritten with expressing x_1 on the left side, the second equation for x_1, and so on.

$$
\begin{aligned}
x_1 &= \frac{\beta_1 - \alpha_{12}x_2 - \alpha_{13}x_3 \ldots - \alpha_{1n}x_n}{\alpha_{11}} \\
x_2 &= \frac{\beta_2 - \alpha_{21}x_1 - \alpha_{23}x_3 \ldots - \alpha_{2n}x_n}{\alpha_{22}} \\
&\vdots \\
x_n &= \frac{\beta_n - \alpha_{n1}x_1 - \alpha_{n2}x_2 \ldots - \alpha_{n,n-1}x_{n-1}}{\alpha_{nn}}
\end{aligned}
\tag{7.9}
$$

These equations can be summarized as follows:

$$
x_n = \frac{\beta_n - \sum_{\substack{j=1 \\ j \neq n}}^{n} \alpha_{nj}x_j}{\alpha_{nn}}
\tag{7.10}
$$

$$
x_i = \frac{\beta_i - \sum_{\substack{j=1 \\ j \neq i}}^{n} \alpha_{ij}x_j}{\alpha_{ii}}
\tag{7.11}
$$

where $i = 1, 2, \ldots, n$.

Now, to get x_i, one makes an initial assumption about the values of x_i and then computes the updated estimates using the modified equations. To construct the following estimates, x_i, one must always utilize the most recent estimates. The unknowns are only saved and used in subsequent calculations with their most recent (r) values. The iteration begins with an initial guess for the unknowns, denoted by x_i^0, and further iterative values are determined, denoted in the equation above by the superscript. When a convergence condition like is met, the iteration comes to an end. One determines the absolute relative approximate inaccuracy for each x_i as at the conclusion of each iteration,

$$
\epsilon_i \geq \left| \frac{x_i^{r+1} - x_i^r}{x_i^r} \right| \text{ for } 1, 2, 3, \ldots, n
\tag{7.12}
$$

Modelling and Simulation of Thermal Energy System 121

Here, ϵ is a convergence parameter chosen such that, if it is reduced further, the results are essentially unaffected. The numerical modelling of thermal systems employs both direct and iterative techniques. Unless they can be acquired in the form of a tridiagonal system, large sets of equations that arise in the finite difference and finite element solutions of partial differential equations are often solved by iterative methods. Direct techniques are more precise and efficient for smaller equation sets. Thermal systems, however, frequently result in nonlinear equations that must be solved iteratively.

Nonlinear algebraic and differential equations are typically produced through the mathematical modelling of thermal systems. This is because most heat and fluid flow conditions are characterized by nonlinear transport mechanisms. Nonlinear equations require substantially more work to solve than linear equations do. Additionally, several solutions might be found, necessitating additional information, especially from the physical nature of the problem, to select the best one. Iterative approaches are required because, with the exception of a few rare circumstances like the quadratic equation, direct solutions to the equations are not feasible. In fact, the linear problem created by linearizing the nonlinear problem is then employed to generate the solution through iteration.

A single nonlinear equation's solution or roots can be found by identifying the values of x that would fulfil the given equation, which may be a polynomial equation of degree n of the type,

$$f(x) = x^n + \alpha_1 x^{n-1} + \alpha_2 x^{n-2} + \ldots + \alpha_{n-1} x + \alpha_n = 0 \tag{7.13}$$

where α represents real coefficients. This equation has n number of real or complex roots involving logarithmic, trigonometric, exponential and other functions. Most of the time, the physical context of the problem under examination might reveal the root's type and the general area in which it lies.

The Newton-Raphson method, which determines the subsequent iterative approximation to the root x_{i+1}, uses the iterative approximation to the root x_i to solve roots, is arguably the most significant and popular technique, and neglecting the higher-order terms derived from the truncated Taylor series expansion, we get

$$f(x) \approx f(x_i) + f'(x_i)(x_{i+1} - x_i) \approx 0 \tag{7.14}$$

which gives an iterative process for finding the root, starting with an initial guess x_1. When the convergence criterion, as stated in equation (7.12), is met, the process is considered complete. The initial guess and the characteristics of the equation could, however, cause the iteration process to diverge. A new initial guess is selected and the process is repeated if the plan deviates.

122 Thermal Energy Systems

Many times, sets of nonlinear equations result from the mathematical modelling of thermal systems. These equations are often solved using iteration and by combining root-finding and linear system-solving techniques. Newton's method and sequential substitution method are two crucial strategies for resolving a set of nonlinear algebraic problems.

If the general system of equations with multiple variables $(x_1, x_2, ..., x_n)$ presented by

$$
\begin{aligned}
f_1(x_1, x_2, x_3 \ ... \ x_i \ ... \ x_n) &= 0 \\
f_2(x_1, x_2, x_3 \ ... \ x_i \ ... \ x_n) &= 0 \\
&\vdots \\
f_j(x_1, x_2, x_3 \ ... \ x_i \ ... \ x_n) &= 0 \\
&\vdots \\
f_m(x_1, x_2, x_3 \ ... \ x_i \ ... \ x_n) &= 0
\end{aligned}
\tag{7.15}
$$

where i represents the number of variables between 1 and n, j represents the equation numbers in the range of 1 and m. Following the Newton-Raphson iterative scheme for the system of equations neglecting the higher-order terms, we can derive,

$$
\begin{bmatrix} \frac{\partial f_1}{\partial x_1} & \cdots & \frac{\partial f_1}{\partial x_n} \\ \vdots & \ddots & \vdots \\ \frac{\partial f_m}{\partial x_1} & \cdots & \frac{\partial f_m}{\partial x_n} \end{bmatrix} \begin{pmatrix} x_{1,k+1} - x_{1,k} \\ \vdots \\ x_{n,k+1} - x_{n,k} \end{pmatrix} = \begin{bmatrix} f_1(x_{1,k}, x_{2,k}, x_{3,k} \ .... \ x_{n,k}) \\ \vdots \\ f_m(x_{1,k}, x_{2,k}, x_{3,k} \ .... \ x_{n,k}) \end{bmatrix}
\tag{7.16}
$$

$$
(\bar{a})(\mathbf{X}) = (\bar{b})
\tag{7.17}
$$

where $(\bar{a})$ is coefficient matrix given by the derivatives of the functions, $(\bar{b})$ is column vector given by the set of function at the previous approximate values, $(\mathbf{X})$ is column vector given by the correction value $\Delta x = x_{n+1} - x_k$, the difference between the present approximate value and previous approximate value, where $x_{1,k}, x_{2,k}, x_{3,k} ...., x_{n,k}$ are estimated values of the unknown variable in the previous iteration, $x_{1,k+1}, x_{2,k+1}, x_{3,k+1} ..., x_{n,k+1}$ are estimated values of the unknown variable in the present iteration. Depending on the nature of the problems, several different function types can be used to analytically determine the partial derivative of all functions. However, the functions frequently cannot be precisely differentiated analytically. The derivates are numerically derived in such instances.

Until the successive iterated values of the approximate solution diverge by less than a set tolerance limit, the operation is repeated with the current or new estimate of the variables, as stated in equation (7.12). If the initial guess is too far off from the precise solution, the plan can diverge.

The choice of the initial guess is frequently guided by the physical nature of the thermal system and previous solutions. The Newton-Raphson method frequently employs an under-relaxation technique for specific variables that may differ significantly during the opening few iteration steps in order to improve convergence rates. The Newton-Raphson technique can be used to solve a system of nonlinear equations by implementing the sequential under-relaxation scheme [10]. There are other popular methods such as Gauss-Seidel method, Gauss elimination method, etc. However, Newton's method generally has better convergence characteristics than the others [11].

For the transient lumped modelling of numerous thermal systems, ordinary differential equations, which involve functions of a single independent variable and their derivatives, are used. An example of a general nth-order differential equation is

$$\frac{d^n y}{dx^n} = f\left(x, y, ..., \frac{dy}{dx}, ..., \frac{d^{n-1} y}{dx^{n-1}}\right) \tag{7.18}$$

For a solution to this problem, n separate boundary conditions are required. The issue is known as an initial-value problem if all of these requirements are stated for a single value of x. A boundary-value problem is one in which the requirements are specified for two or more values of x [12]. A simple first-order differential equation is

$$\frac{dy}{dx} = f(x, y) \tag{7.19}$$

with the boundary condition

$$y(x = x_0) = y_0 \tag{7.20}$$

To solve this differential equation numerically, one must determine the value of the function $y_k(x)$ at discrete values of x, such as

$$x_k = x_0 \pm k\Delta x \text{ with } k = 1, \ 2, \ ... \tag{7.21}$$

Runge-Kutta method is mostly used for determining the values of $y_1, y_2, y_3,$ ... for the dependent variable $y_k(x)$ corresponding to these discrete values of x, in such initial value problems. The fourth-order Runge-Kutta techniques are self-contained, reliable, and easy to apply. As a result, they are widely used and most computers have the necessary software. The conventional approach is described in,

124 Thermal Energy Systems

$$y_{k+1} = y_k + \Delta x \left\langle \frac{R_1}{6} + \frac{R_2}{3} + \frac{R_2}{3} + \frac{R_4}{6} \right\rangle$$

$$R_1 = f(x_k, y_k)$$

$$R_2 = f\left(x_k + \frac{\Delta x}{2}, y_k + \frac{\Delta x}{2}R_1\right) \tag{7.22}$$

$$R_3 = f\left(x_k + \frac{\Delta x}{2}, y_k + \frac{\Delta x}{2}R_2\right)$$

$$R_4 = f\left(x_k + \Delta x, y_k + \Delta x R_3\right)$$

The total error varies as $(\Delta x)^4$, indicating that it is a fourth-order system. Higher-order (>4) Runge-Kutta techniques are not very effective. Higher-order approaches do not appear to have precise data available [11]. If we wish to build a Runge-Kutta method that adaptively selects the step size for the time step in order to minimize the local truncation error, some higher-order methods may still be useful. As indicated earlier, a system of first-order equations is solved in order to solve higher-order equations. To determine the values of all the unknowns at the following stage, the computations are completed in order. To utilize this strategy, all the prerequisites in terms of y and its derivatives must be known at the outset. As a result, the technique as it is presented here works for initial-value problems [13].

We typically deal with issues where the boundary conditions are presented at two or more different values of the independent variable when simulating thermal systems. Boundary-value problems are a category of such issues. The differential equation must at least be of second order in order to give rise to a boundary-value problem where the two conditions are given at two separate values of the independent variable since the number of boundary conditions required equals the order of the differential equation. Take the second-order equation shown below as an illustration:

$$\frac{d^2y}{dx^2} = f\left(x, y, \frac{dy}{dx}\right) \tag{7.23}$$

with the boundary conditions

$$y = L \text{ at } x = l, \text{ and } y = M \text{ at } x = m$$

By using the first boundary condition and assuming a presumed value of the derivative at, say, $x = 1$ for the previous problem, you can first reduce the problem to an initial-value problem. This derivative is adjusted by iteration so that it also satisfies the boundary condition at $x = m$. The rectification approach uses root-solving methods like the Newton-Raphson method. The adjustment of beginning conditions to satisfy the conditions at the other place are analogous to shooting at a target, hence procedure-based solutions is known as "shooting methods" [3]. The method is easily adaptable to higher-order equations and various boundary conditions.

Modelling and Simulation of Thermal Energy System 125

The situation where the thermal parameters are functions of space and may be also of time is one that occurs frequently in the numerical modelling of thermal systems. The differential equations that control such issues require partial derivatives and are referred to as partial differential equations if the dependent variable is a function of two or more independent variables [14]. The common type of partial differential equations that appear frequently in thermal systems:

$$\frac{1}{\alpha}\frac{\partial T}{\partial t} = \frac{\partial^2 T}{\partial x^2} \text{ with } x > 0, \quad t > 0, \quad \alpha = \kappa/\varrho C \tag{7.24}$$

where T is the temperature, x is the coordinate axis, t is the time, and $\alpha = \kappa/\varrho C$, is the thermal diffusivity of the material with κ, C, ϱ, thermal conductivity, specific heat, and density of each layer, respectively. A demonstrating in time t solution can be used to solve the first one, which is a parabolic equation. It needs an initial time condition, and two x boundary conditions. The second equation is an elliptic equation, which necessitates correctly stated conditions on the domain's entire boundary. There are several numerical solutions to partial differential equations that appear in fluid flow and heat transfer systems. The two primary strategies, the finite difference and the finite element methods, are just briefly described here.

7.2.4.1.1 Finite Difference Method

As shown in Figure 7.4, finite difference method involves imposing a grid on the computing domain to produce a finite number of grid points. The values at these grid locations are used to define the partial derivatives in the specified partial differential equation. The discretized forms of the various derivatives are often derived via Taylor series expansions. These result in a set of algebraic equations that are represented as finite difference equations for each grid point. The discrete points on the grid, which are known as the mesh's nodes, serve as a replacement for the continuous fixed solution region in the finite difference technique. The discrete variable functions created on the mesh are then estimated to replace the continuous fixed solution region. The difference quotient roughly represents the relationship between the continuous variable and the solution region, the derivative of the original equation, and the fixed solution condition. Then, from the discrete solution, one can approximate the solution in the entire region using the interpolation method.

Equation (7.24) can be written in finite difference form as

$$\frac{T_{i+1,j} - T_{i,j}}{\Delta t} = \alpha_k \left\{ \frac{T_{i,j+1} - 2T_{i,j} + T_{i,j-1}}{(\Delta x)^2} \right\} k = 1, 2, 3, 4 \tag{7.26}$$

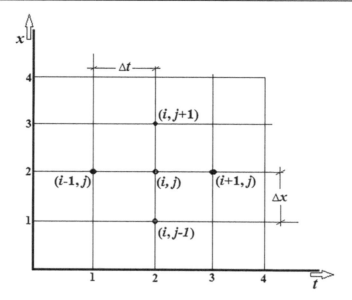

Figure 7.4 A two-dimensional finite difference method grid diagram.

where the subscript $(i + 1)$ and i denote the nodal values at time $(t + \Delta t)$ and t, respectively. The spatial location, correspondingly, is given by j, so, $x = j\Delta x$ and $t = i\Delta t$. For this approximation, the truncation error, which represents the error resulting from terms missed in the Taylor series, has an order of Δt in time and $(\Delta x)^2$ in space. At time t, the first derivative is taken as a forward difference, and the second derivative is approximately calculated. What follows is a finite difference equation as

$$T_{i+1,j} = T_{i,j} + \mathscr{U}\alpha_k \{T_{i,j+1} - 2T_{i,j} + T_{i,j-1}\} \text{ with } \mathscr{U} = \Delta t/(\Delta x)^2, \; k = 1, 2, 3, 4 \quad (7.27)$$

This equation gives the temperature distribution at time $(t + \Delta t)$ at the grid point whose spatial coordinate is $x = j\Delta x$, in terms of temperatures at time t at the grid points with coordinates $(x - \Delta x)$, x and $(x + \Delta x)$. The temperature distribution can be calculated for increasing values of time t if the starting temperature distribution and the boundary conditions are known. This is the explicit approach, which is often referred to as the forward time central space approach. However, the numerical scheme's stability can only be guaranteed if $\mathscr{F} = \alpha \mathscr{U} \leq 0.5$, where $\mathscr{F}$ is known as the grid Fourier number. The coefficients in equation (7.27) must all be positive for the scheme to be stable, which is ensured by this limit of $\mathscr{F}$. The procedure is hence conditionally stable [14]. This issue can be addressed using the same techniques to the multidimensional issues thermal systems frequently faced.

Modelling and Simulation of Thermal Energy System 127

For instance, the following equation governs two-dimensional, unstable conduction at constant properties:

$$\frac{1}{\alpha}\frac{\partial T}{\partial t} = \frac{\partial^2 T}{\partial x^2} + \frac{\partial^2 T}{\partial y^2} \tag{7.28}$$

Stability considerations again pose a limitation of the form $\mathcal{F} \leq 0.25$, if $\Delta x = \Delta y$. For further details, Patankar [15] and Tannehill et al. [16] may be consulted.

7.2.4.1.2 Finite Element Method

The integral formulation of the conservation principles serves as the foundation for the finite element approach. A collection of finite elements, available in various types and forms for various geometries and governing equations, is used to describe a computational area of interest. The nodal values of a sought-after physical field are used to determine approximating functions in finite elements. Discrete finite element problems with unknown nodal values are created from continuous physical problems. The most prevalent types of components are linear elements for one-dimensional applications, triangular elements for two-dimensional issues, and tetrahedral elements for three-dimensional issues. The dependent variable's variation is typically assumed to be a polynomial and frequently to be linear within the elements. The conservation principles are satisfied by minimizing the integrals or by bringing their residuals to zero for the integral equations that apply to each element. Galerkin's approach [16] is a weighted residual method that is frequently used for thermal processes and systems. The variational formulation of the finite element equations is typically utilized if the physical problem can be expressed as the minimization of a function.

A discrete model of the problem should be supplied in numerical form in order to operate finite element methods. A solution region is initially divided into finite elements. A preprocessor software normally creates the finite element mesh. Nodal coordinates and element connectivities are the two main arrays used in the mesh description. The variables from the field are interpolated over the element using interpolation methods. Polynomial functions are frequently chosen as interpolation functions. The polynomial's degree is determined by how many nodes are assigned to each constituent. It is necessary to establish the matrix equation for the finite element that connects the nodal values of the unknown function to other variables. We must put together every element equation in order to find the global equation system for the entire solution region. In other words, all local element equations for elements utilized in discretization must be combined. During assembly, element connectivities are employed.

128   Thermal Energy Systems

Boundary conditions should be enforced before solving because they are not taken into account in element equations. The global equation system for finite elements is often positive definite, symmetric and sparse. There are direct and iterative approaches for solving problems. The result of the solution is the nodal values of the requested function. We frequently need to calculate extra parameters. For instance, in mechanical problems, strains and stresses, as well as displacements, which are produced after solving the system of global equations, are of interest.

For practical analysis, finite element models can include tens of thousands or even hundreds of thousands of degrees of freedom. These meshes cannot be produced manually. A software tool known as a mesh generator separates the solution domain into several finite element subdomains. Mesh generators come in a variety of forms. We wish to specifically discuss two types for two-dimensional issues: block mesh generators and triangulators. Block mesh generators require some kind of large partitioning at the beginning. The solution domain is divided into a select few relatively small chunks. There should be a uniform form for each block. Typically, mapping approach is used to create the mesh inside the block. Typically, triangulators create irregular meshes inside of arbitrary domains. Mesh generation frequently uses Delaunay triangulation and Voronoi polygons. Later, a triangle mesh can be changed into a quadrilateral element mesh. It is possible to generalize Delaunay triangulation to three-dimensional domains [17].

Let's say that we wish to create a quadrilateral mesh inside a domain with a curved quadrilateral shape. This can be accomplished by using Figure 7.5. If a parabola can be used to approximate each side of the curved quadrilateral domain, then the domain resembles an 8-node isoparametric element. The domain is mapped to a square in the local coordinate system ξ, η. The nodal coordinates are converted back to the global coordinate system of x, y after the square in coordinates is partitioned into rectangular elements. Algorithm of coordinate calculation for node i:

n_ξ = number of elements in ξ direction, n_η = number of elements in η direction

$$\text{Row: R} = \frac{(i-1)}{\left(n_\xi + 1\right)} + 1 \quad \text{Column: C} = \text{mod}\left|(i-1), \left(n_\xi + 1\right)\right| + 1 \quad (7.29)$$

$$\Delta\xi = 2/n_\xi, \eta = 2/n_\eta$$
$$\xi = -1 + \Delta\xi(C-1)$$
$$\eta = -1 + \Delta\eta(R-1)$$
$$x = \Sigma N_P(\xi, \eta)x_P$$
$$y = \Sigma N_P(\xi, \eta)y_P$$

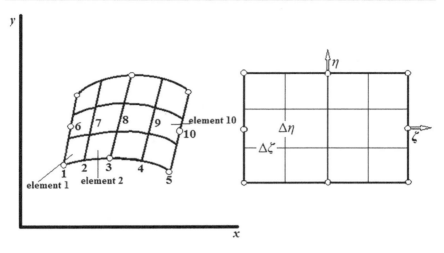

Figure 7.5 Mesh generation with mapping technique.

Connectivity for element e:

$$\text{Element row: } R = \frac{(e-1)}{n_\xi} + 1$$

$$\text{Element column: } C = \text{mod}\left|(e-1), n_\xi\right| + 1$$

Connectivity (global node numbers):

$$\begin{aligned}
i_1 &= (R-1)(n_\xi + 1) + C \\
i_2 &= i_1 + 1 \\
i_3 &= R(n_\xi + 1) + C + 1 \\
i_4 &= i_3 - 1
\end{aligned}$$

To refine (create smaller elements) the mesh near a domain corner, mid-side nodes can be moved closer to the corner. The mid-side node should be shifted to the following location if refining is performed on the element side that is parallel to the local axis and the size of the smallest element close to the corner node is Δd:

$$\varepsilon = \frac{d_m}{d} = \frac{1 + \frac{\Delta d}{d} n_\xi - \frac{2}{n_\xi}}{4\left(1 - \frac{1}{n_\xi}\right)} \tag{7.30}$$

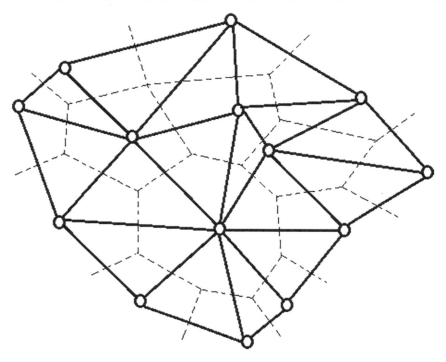

Figure 7.6 Mesh generation in Delaunay triangulation and Voronoi polygons.

Here, d_m is the length between the corner and mid-side nodes; and d is the length of the element's side.

Dashed lines in Figure 7.6 depict voronoi polygons. Three nearby polygons share a vertex of a Voronoi polygon. A triangle is formed by joining three points connected to three such neighbouring polygons. The Delaunay triangulation refers to the collection of triangles. A quadrilateral element-based mesh can be created using a generated triangle mesh. For this reason, connecting triangles with subsequent quality enhancement is typically used.

A set of algebraic equations is the end result of using the finite element method in the computational domain. The contributions from each element are combined to create the overall set of equations, or global equations. Condensation is a method used to eliminate interior nodes from the built system. When the set of equations is solved, the values at the nodes are produced, and the interpolation functions are then used to produce values across the entire domain. By selecting and placing finite elements correctly, the approach can handle complex geometries. Variations in material properties and arbitrary boundary constraints can be simply implemented. The strategy is particularly flexible because it may be used to a variety of issues. Finite element techniques can adapt to arbitrary changes in boundaries and characteristics. As a result, the finite element method has formed the foundation for a large portion of the software created for engineering systems and processes over the past two decades. For simpler geometries and boundary conditions, finite difference methods remain common.

7.2.5 Physical Models

A model that resembles the real system in terms of shape, geometry, and other physical features is referred to as a physical model. Scale models that are smaller than the full-size system are of particular importance in design since experimenting on a full-size prototype is either impractical or prohibitively expensive. The model might potentially be a condensed version of the real system or it could concentrate on a single element of the system. These models are subjected to experiments, and the outcomes are used to represent the behaviour and traits of the relevant component or system. As a result, the data from physical modelling serves as both data for the mathematical model's validation as well as inputs for the design process. Due to the intricacy of the transport mechanisms that develop in typical actual systems, physical modelling is particularly crucial for designing thermal systems. Many times, mathematical modelling cannot be used to appropriately reduce the issue and produce an accurate solution that closely resembles the real system. Additionally, some of the estimations may not be quite accurate. Then, experimental data are required to evaluate the validity and accuracy of the model [18]. The fundamental mechanisms are sometimes difficult to model. Then, for a suitable description of the problem, experimental inputs are required. However, experimental work takes a lot of time and money. Therefore, it's important to reduce the number of tests required to gather the desired data. The scope of this text prohibits further examination of the subject. To learn more, readers might examine other books.

7.3 INTEGRATION OF DIFFERENT MODELS

The different numerical models for the many components of the specified thermal system must all be generated, tested, and combined to create the model for the entire system based on physical reasoning and analytical findings. The actual boundary condition must be used in place of the approximations when the two models are combined. Through the governing equations shared by these parts at a particular time, the boundary parameters at the intake of the succeeding component are made equal to those to the equivalent values at the exit of the previous component for the overall thermal system. As you continue in this manner, additional components become related via the boundary and inflow/outflow conditions. It may seem like an overly complex method to develop the numerical model

132 Thermal Energy Systems

for the system by first modelling the individual pieces and then coupling them. In fact, it is frequently more practical and effective to create the numerical model for the system without taking individual pieces into consideration for relatively basic systems with few parts. However, if the system consists of several components, it is best to create discrete numerical models, test and validate them independently, and then combine them to create the system model. This makes it possible to divide a challenging issue into smaller ones that may be handled and evaluated separately before being finally assembled. In industry, this method is frequently used to represent complicated systems. There are numerous coupled equations involved; hence there are few chances for success in a direct simulation of the complete system.

7.4 SYSTEM SIMULATION

System simulation is the process of using an analysis, research, or examination of a system model to gather quantitative data on the behaviour and characteristics of the real system. A system simulation is similar to seeing a manufactured system that replicates the functionality of an actual system. In order to give the quantitative inputs required for design and optimization, the governing equations generated from the mathematical model are solved by analytical or numerical methods to provide the system performance of the system under various operating situations as well as for different design factors. To ascertain the predicted operating circumstances and performance at design as well as off-load or part-load conditions at all operational points of the system, system simulations are carried out. This aids in identifying the variables' limiting circumstances. All operational variables, including mass flow rates, temperature, pressure, density and other variables present throughout the system, are calculated during a system simulation. It assumes knowledge of all mass and energy balance equations, constitutive relations as well as equations for the thermodynamic properties of the working substances, and performance characteristics of all components. A set of simultaneous equations is made up of the thermodynamic properties, equations for how well each component performs, and equations for the conservation of mass and energy. This can be applied to an existing design to investigate the possibility of performance-improving alterations, while it is typically used in the design stage to examine alternatives and/or an improved design. One of the most crucial components in the design and development of thermal systems is system simulation. We must rely on simulation based on a model of the supplied system to gain the needed information on the system's behaviour under various conditions because experimenting on a prototype of the actual thermal system is typically quite expensive and time-consuming. Another significant application is the use of simulation to fix a flaw in an existing system or to alter the system to increase performance. Instead of altering a specific component to fix the issue or enhance the system, simulation is first utilized to ascertain the impact of such a modification. Since the simulation closely mimics the real physical system, it is possible to assess the appropriateness of the suggested adjustment without actually making it.

Thermal systems use a variety of simulation techniques. While numerical modelling and numerical simulation are based on the mathematical modelling of the thermal system, analogue and physical simulations are based on the equivalent kind of modelling. The main three classifications of simulation are discussed next.

7.4.1 Steady or Dynamic Simulation

Situations, where changes with respect to time are small or nonexistent, are referred to as steady-state simulations. Dynamics simulations are performed for systems where one of the parameters evolves over time and are time dependent. Due to the time-dependent nature of the majority of thermal systems, a dynamic simulation is crucial. This is especially accurate for systems put in place and shutdown. A steady-state simulation is substantially easier than the related dynamic simulation since the variables' dependence on time is removed. Steady-state simulation is more interesting and significant in thermal systems due to the fact that the steady-state approximation can be used in a large number of practical scenarios. Most systems operate as though they are in steady-state settings, with the exception of moments just before start-up and discontinue. Additionally, dynamic simulation is required to investigate how the system reacts to abrupt changes in operating conditions. The outcomes of a dynamic simulation can also be used to build and research a control strategy for the system's successful operation.

7.4.2 Continuous or Discrete Simulation

It is assumed that the substance or fluid is a continuum, and the conservation rules are obtained using a continuum approximation. This indicates that for transport processes in a continuum, continuity, momentum, and energy equations from fluid mechanics and heat transfer are used. If particles are present in the flow, they are not considered independently but rather as a part of the fluid's typical features. Conversely, if discrete parts are taken into account, the simulation concentrates on a limited number of these elements. In these situations, each item's mass, momentum and energy balances are taken into account separately to determine. The resolution of differential equations enables continuous simulation. Contrarily, discrete simulation is used in systems where the quantities and states change only at discrete moments in time.

134 Thermal Energy Systems

7.4.3 Deterministic or Stochastic Simulation

Most thermal system examples presume that the variables in the equation are precisely stated. These procedures are said to as deterministic. A probability distribution with a dominating frequency, an average, and amplitude of variation may be provided in place of precise input circumstances in a number of situations. The right design variables are chosen by using probabilistic descriptions, and the corresponding simulation, known as stochastic, which can also be used when the conditions are fully random and have an equal chance of achieving any value within a certain range.

7.5 METHODOLOGY

The best approach for modelling a specific thermal system depends heavily on the nature of the system and, consequently, the features of the governing equations. The mathematical model serves as the finest framework for describing these.

Information flow between the many sections, components, or subsystems that make up the system is a valuable idea in the simulation of thermal systems. Information is sent from one component to another primarily in terms of physical properties that are particularly relevant to thermal systems, such as temperature, velocity, flow rate, and pressure, and this information reveals how the two are coupled. The system parts may supply the input to a particular object undergoing thermal processing or to a continuous flow, and conversely, they may receive the output from this item or flow. This involves the movement of mass, velocity, and energy through the system as discrete objects or a continuous flow. Obviously, depending on the thermal system, many information flow arrangements could exist. The many components of the system are connected through inputs and outputs in such arrangements, which are frequently represented as information-flow diagrams.

A system's information flow diagram, which depicts the flow of all input and output variables, is made up of block diagrams for each key component. Each system component is represented by a block diagram, and the performance characteristics are expressed by a transfer function. The information-flow diagram for a vapour-compression cooling system as an illustration of one such block diagram is provided in Figure 7.7. In this closed cycle, the entry into each component corresponds to the exit from the one before it. The graphics show the relationship between various system components and hence offer suggestions for how to approach the simulation. The characteristics of the compressor, condenser, throttling valve and evaporator can be expressed by the following equations [3]:

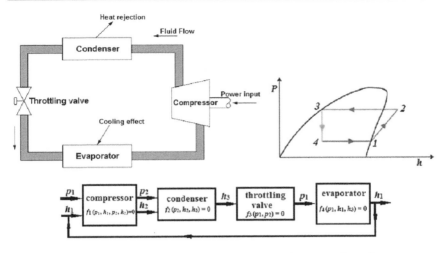

Figure 7.7 Illustration of a vapour-compression system information-flow diagram.

$$f_1(p_1, h_1, p_2, h_2) = 0$$
$$f_2(p_2, h_2, h_3) = 0$$
$$f_3(p_1, p_2) = 0 \qquad (7.31)$$
$$f_4(p_1, h_1, h_3) = 0$$

7.6 METHODS FOR NUMERICAL SIMULATION

The accuracy of the simulation can typically be maintained by modelling the majority of thermal and fluid flow systems as stable and lumped, including network pipes, power plants, air conditioners, internal combustion engines, gas turbines, compressors, pumps, etc. [19]. If all of the equations are linear, the system of equations can be solved directly or iteratively. The former method is used for tridiagonal systems and relatively small sets of equations, while the latter method is used for large sets, as was stated earlier.

The sequential substitution and Newton-Raphson methods, which were previously mentioned for a single nonlinear algebraic equation and for a collection of nonlinear equations, are the two basic methodologies used for simulating thermal systems governed by nonlinear equations. Each equation in a system of equations is solved for unknowns using previously iterated calculations' known results. The following equation is then solved to determine another unknown using this solution as a substitution. This is successively added back into the following equation, and the process is repeated until the result does not noticeably change from one iteration to the next. To effectively model the system, a method like the modified Gauss-Seidel method may be utilized. The difficulty in getting the iterative process to converge is the fundamental issue with the successive substitution method. The Newton-Raphson method for a single nonlinear problem and then extended to a set of nonlinear algebraic equations is the second method for solving a set of nonlinear algebraic equations. This approach is excellent for an information-flow diagram if there is a lot of inter-dependence between the various system components. The derivatives used in the Newton-Raphson technique connect all the system's components concurrently. As a result, modifications to one component will have an immediate impact on all the others, with the outcome depending on the related derivatives. As a result, the strategy mimics the physical behaviour of systems where all of the components are intimately connected.

Despite the fact that the assumption of lumping or uniform conditions in each system component has been, and continues to be, very popular due to its simplicity, it has become quite simple to model and simulate the more general distributed circumstance in which the quantities vary with location and time. There are numerous actual cases where the domain has significant changes and the lumped approximation cannot be applied. Partial differential equations, which are frequently nonlinear due to changes in material properties, coupling with fluid flow, and the presence of radiative transport, are the governing equations for dispersed systems. The methods for extracting simultaneous algebraic equations from the governing partial differential equations and solving them using finite difference and finite element methods were described.

The presence of a high number of components, which results in huge sets of governing equations, and the existence of generally independent subsystems that make up the entire system are the two fundamental characteristics that set large thermal systems apart from simpler ones. Large sets of algebraic equations can be handled with the minimum amount of computer storage by using techniques like Gauss-Seidel. Similar to this, the modularization of the simulation process has been emphasized in various places since it enables the expansion of the system simulation package while guaranteeing the adequate treatment of each subsystem. The system simulation is used to obtain the numerical inputs required for design and optimization, for studying off-design conditions, for establishing safety limits, and for examining the sensitivity of the system to various design parameters after it has been thoroughly validated and the precision of its predictions determined [20,21].

7.7 OPTIMIZATION OF THERMAL SYSTEMS

It is assumed that a system that has been constructed or put together thanks to a design will carry out the proper functions for which the effort was made. However, as the optimal design is determined by factors like cost, performance, efficiency and other factors, the design would typically not be the greatest design. In actual fact, our goals are typically to achieve the best quality or performance for the money while minimizing any negative environmental effects. This introduces the idea of optimization, which reduces or enhances quantities and properties that are specifically important to a given application. The process of optimization involves choosing a combination of variables or parameters those results in the maximum or minimum value of a function that describes the system's performance, or the optimum design. Various combinations of the key factors and variables are tried out until the design is optimized [22–25]. Under certain restrictions, the maximum and lowest values of a representative analysis function, often a cost function or a performance function, are typically sought. Creating a functional system that accomplishes the required objective while adhering to the limitations set by safety, environmental, economic and other similar concerns is no longer sufficient. It has become crucial to optimize the process in order to maximize or minimize a selected variable because of the escalating global competitiveness and the requirement to boost efficiency. The most common name for these variables is the objective functions [3]. In design, optimization is critical and has grown even more so in recent years as a result of increased worldwide competition. Having a functional system that completes the required tasks and adheres to the established limitations is no longer sufficient. The final design, which minimizes or maximizes a suitably specified quantity, should be picked from at least a few workable designs that were produced. The cost and performance of a system are generally influenced by a wide range of factors. Designers can assess a variety of alternative system solutions through an iterative process based on the system's cost-effectiveness, efficiency, dependability and durability [26–30].

Energy and fluid flow are the primary concerns in thermal systems. As a result, the goal function is commonly based on energy consumption, which takes into account factors like fuel consumption, energy exchange with the environment, efficiency of the system and its components, energy transit and losses, etc. The rate of energy consumption per unit output, where the output can be power given, heat eliminated, goods made, etc., is a valuable objective function. The best design is then the one with the lowest energy use per unit of output. The system that produces the most output for a given amount of energy is optimal. This objective function can alternatively be thought of as the output per unit cost because energy use can be stated as a cost. Efficiency, output per unit fuel consumption, output per unit cost, heat transfer rate, and fluid flow rate are some of the physical quantities that are frequently maximized in thermal systems, whereas energy losses, energy input, pressure head for fluid flow, fluid leakage, rate of fuel consumption per unit output, etc., are often minimized [31–33].

The constraints in thermal systems are primarily caused by the conservation laws for mass, momentum, and energy, as well as by earlier described restrictions on the type of material, available space, and

138 Thermal Energy Systems

equipment. However, in typical systems of practical interest, these typically result in nonlinear numerous partial differential equations with complex geometries and boundary conditions. Due to the properties of the materials, combined thermal transport methods, etc., additional complexity may also develop. The fundamental issue created by these complexities is that it takes a lot of time and effort to simulate the system for each set of conditions. As a result, it is typically necessary to reduce the quantity of simulations required for optimization. With the aid of curve fitting, numerical or experimental simulation findings for relatively basic thermal systems can be utilized to derive algebraic expressions and equations that describe the system's behaviour. The optimization problem then becomes simple, and the optimum can be obtained using a variety of methods. Unfortunately, only a few straightforward and frequently unworkable situations allow for the use of this strategy. Numerical modelling is used to obtain the simulation results for common practical systems in order to achieve the optimal. If a prototype is available, experimental data are also employed, but again, the availability of such data is limited because experimental runs are typically costly and time-consuming [34–36].

REFERENCES

1. Doebelin, E. O., *System Modeling and Response*, Wiley, New York, 1980.
2. Szucs, E., *Similitude and Modeling*, Elsevier, New York, 1977.
3. Jaluria, J., *Design and Optimization of Thermal Systems*, 2nd ed., CRC Press Taylor & Francis Group, New York, 2008.
4. Incropera, F. P., Dewitt, D. P., *Fundamentals of Heat and Mass Transfer*, 5th ed., Wiley, New York, 2001.
5. Eckert, E. R. G., Drake, R. M., *Analysis of Heat and Mass Transfer*, McGraw- Hill, New York, 1972.
6. Bejan, A., *Heat Transfer*, Wiley, New York, 1993.
7. Cengel, Y. A., Boles, M. A., *Thermodynamics: An Engineering Approach*, 4th ed., McGraw-Hill, New York, 2002.
8. Rieder, W. G., Busby, H. R., *Introductory Engineering Modeling Emphasizing Differential Models and Computer Simulation*, Wiley, New York, 1986.
9. Hilderbrand, F. B., *Methods of Applied Mathematics*, Prentice-Hall, Englewood Cliffs, NJ, 1965.
10. Majumdar, P., *Computational Methods for Heat and Mass Transfer*, Taylor & Francis, New York, 2005.
11. James, M. L., Smith, G. M., Wolford, J. C., *Applied Numerical Methods for Digital Computation*, 3e. Harper Collins Publishers, New York, NY, 1985.
12. Gerald, C. F., Wheatley, P. O., *Applied Numerical Analysis*, 5th ed., Addison-Wesley, Reading, MA, 1994.
13. Ivrii, V., *Partial Differential Equations*, Department of Mathematics, University of Toronto, Canada, 2021, http://www.math.toronto.edu/ivrii/PDE-textbook/PDE-textbook.pdf.

14. Smith, G. D., *Numerical Solution of Partial Differential Equations*, Oxford University Press, Oxford, U.K., 1965.
15. Patankar, S. V., *Numerical Heat Transfer and Fluid Flow*, Taylor & Francis, Washington, DC, 1980.
16. Tannehill, J. C., Anderson, D. A., Pletcher, R. B., *Computational Fluid Mechanics and Heat Transfer*, 2nd ed., Taylor & Francis, Washington, DC, 1997.
17. Reddy, J. N., *An Introduction to the Finite Element Method*, 3rd ed., McGraw-Hill, New York, 2004.
18. Wellstead, P. E., *Introduction to Physical System Modeling*, Academic Press, New York, 1979.
19. Howell, J. R., Buckius, R. O., *Fundamentals of Engineering Thermodynamics*, 2nd ed., McGraw-Hill, New York, 1992.
20. Bejan, A., Tsatsaronis, G., Moran, M., *Thermal Design and Optimization*, Wiley, New York, 1996.
21. Fox, R. L., *Optimization Methods for Engineering Design*, Addison-Wesley, Reading, MA, 1971.
22. Dutt, N., Binjola, A., Hedau, A. J., Kumar, A., Singh, V. P., Meena, C. S. (2022). Comparison of CFD Results of Smooth Air Duct with Experimental and Available Equations in Literature, *International Journal of Energy Resources Applications (IJERA)*, Vol. 1, Issue 1, page 40–47, 10.56896/IJERA.2022.1.1.006.
23. Singh, V. P., Jain, S., Kumar, A. (2022). Establishment of Correlations for the Thermo-Hydraulic Parameters Due to Perforation in a Multi-V Rib Roughened Single Pass Solar Air Heater. *Experimental Heat Transfer*, 10.1080/08916152.2022.2064940.
24. Kushwaha, P. K., Sharma, N. K., Kumar, A., Meena, C. S. (2023). Recent Advancements in Augmentation of Solar Water Heaters using Nanocomposites with PCM: Past, Present & Future. *Buildings*, Vol. 13, page 79. 10.3390/buildings13010079.
25. Singh, V. P., Jain, S., Karn, A., Dwivedi, G., Alam, T., Kumar, A. (2022).Experimental Assessment of Variation in Open Area Ratio on Thermohydraulic Performance of Parallel Flow Solar Air Heater. *Arabian Journal for Science and Engineering*, 10.1007/s13369-022-07525-7.
26. Patil, P. P., Kumar, A. (2017). Design and FEA Simulation of Omega Type Coriolis Mass Flow Sensor. *International Journal of Control Theory and Applications*, Vol. 9, Issue 40, page 383–387.
27. Rana, S., Kumar, A. (2018). FEA Based Design and Thermal Contact Conductance Analysis of Steel and Al Rough Surfaces. *International Journal of Applied Engineering Research (IJAER)*, Vol. 13, Issue 16 (2019), page 12715–12724, ISSN 0973-4562.
28. Patil, P., Kumar, A., Gori, Y. (2019), Coriolis Mass Flow Sensor (CMFS): A Review. *Journal of Critical Reviews*, Vol. 6, Issue 6, page 2628–2632, ISSN-2394-5125.
29. Patil, P. P., Kumar, A. (2017). Finite Element Analysis of Omega Type Coriolis Mass Flow Sensor (CMFS) for Evaluation of Fundamental Frequency and Mode Shape. *International Journal of Engineering and Technology*, Vol. 9, Issue 3 S, 10.21817/ijet/2017/v9i3/170903S017.

30. Singh, V. P., Dwivedi, A., Karn, A., Kumar, A., Singh, S., Srivastava, S., Srivastava, K. Nanomanufacturing and Design of High-Performance Piezoelectric Nanogenerator for Energy Harvesting. *Nanomanufacturing and Nanomaterials Design: Principles and Applications*. CRC Press, Taylor & Francis, 2022. ISBN: 9781003220602, 10.1201/9781003220602-15.
31. Meena, C. S., Prajapati, A. N., Kumar, A., Kumar, M. (2022). Utilization of Solar Energy for Water Heating Application to Improve Building Energy Efficiency: An Experimental Study. *Buildings*, Vol. 12, page 2166, 10.3390/buildings12122166.
32. Singh, V. P., Jain, S., Karn, A., Kumar, A., Dwivedi, G., Meena, C. S., Dutt, N., Ghosh, A. (2022). Recent Developments and Advancements in Solar Air Heaters: A Detailed Review. *Sustainability*, Vol. 14, page 12149, 10.3390/su141912149.
33. Singh, V. P., Jain, S., Karn, A., Kumar, A., Dwivedi, G., Meena, C. S., Cozzolino, R. (2022). Mathematical Modeling of Efficiency Evaluation of Double-Pass Parallel Flow Solar Air Heater. *Sustainability*, Vol. 14, page 10535, 10.3390/su141710535.
34. Meena, C. S., Kumar, A., Jain, S., Rehman, A. U., Mishra, S., Sharma, N. K., Bajaj, M., Shafiq, M., Eldin, E. T. (2022). Innovation in Green Building Sector for Sustainable Future. *Energies*, Vol. 15, page 6631, 10.3390/en15186631.
35. Singh, V. P., Jain, S., Karn, A., Dwivedi, G., Kumar, A., Mishra, S., Sharma, N. K., Bajaj, M., Zawbaa, H. M., Kamel, S. (2022). Heat transfer and friction factor correlations development for double pass solar air heater artificially roughened with perforated multi-V ribs. *Case Studies in Thermal Engineering*, Vol. 39, page 102461, ISSN 2214-157X, 10.1016/j.csite.2022.102461.
36. Meena, C. S., Kumar, A., Roy, S., Cannavale, A., Ghosh, A. (2022). Review on Boiling Heat Transfer Enhancement Techniques. *Energies*, Vol. 15, page 5759. 10.3390/en15155759.

Chapter 8

Thermal Efficiency Enhancement of Solar Still Using Fins with PCM

Naveen Sharma, Noushad Shaik, Vivek Kumar, and Mukesh Kumar

CONTENTS

8.1 Introduction 141
8.2 Experimental Facility and Instrumentation 143
8.3 Thermal and Economic Analysis 145
8.4 Results and Discussion 147
8.5 Conclusions 151
References 152

8.1 INTRODUCTION

Renewable energy resources can address many pressing issues facing humanity, including climate change, freshwater shortage, environmental destruction and energy scarcity, and their efficient usage can result in a more sustainable world. Solar energy is a clean, low-cost and abundant source of energy, and is effectively applied to numerous applications including electricity generation, desalination/distillation, air and water heating, cooking and drying [1–5]. Worldwide, especially in remote areas where freshwater supplies are scarce, the demand for fresh water is steadily increasing because of growing populations and industrialization. A major challenge is the provision of fresh water to fulfill the shortage as available groundwater and seawater are unsuitable for household needs, such as cooking and drinking. Water purification systems made from solar technology are a good fit for North India, which receives a high level of solar radiation. Solar stills are currently the most popular method for producing fresh water due to their low fabrication and maintenance costs. The freshwater production rate for plain absorbing plates is very low. As a result, a variety of methods have been proposed to enhance the efficiency and yield of solar stills by modifying their design or using passive/active methods, or heat storage materials (sensitive/latent heat storage) [6–9].

A variety of heat storage materials (HSMs) have been studied by the investigators to boost the yield of solar still during the day and at night,

DOI: 10.1201/9781003395768-8

including cotton and jute cloth, sandstone, clay pots, wicks, sandbags, marble stones and sawdust [10–16]. Sandstone is an effective medium for storing thermal energy owing to its high specific heat capacity and pore holes. This can enhance both daytime and nighttime solar still yields [10]. With 24-hour distillation, solar stills containing sandstone and marble pieces will provide about 30% and 13.6% more accumulation than smooth solar still (SSS), respectively. Experimentally, Sakthivel and Arjunan [11] tested the effect of cotton cloth in one-slope solar still for climatic conditions of Chennai, India. The researchers concluded that solar still with a 6-mm thickness of cotton cloth improved the yield by 21.4% than SSS. Modi and Modi [12] studied adding HSMs such as jute and cotton cloth in two-basin solar still. Experimental findings show that jute cloth produces 21.46% more distilled water when compared with a black cotton cloth. Shehata et al. [13] conducted experiments in solar still with HSM and reported that the highest daily freshwater yield is found to be 5.34 and 7.4 L/day at the water depths of 2.5 and 3.5 cm, respectively.

Yarramsetty et al. [14] assessed the performance of PSS with clay pots facing upward and downward under the weather situations in Andhra Pradesh, India. Experimental results show that the freshwater output obtained for upward-facing clay pots is about 8.6% higher than that of upward-facing clay pots and about 60% higher than that of SSS. Singh et al. [15] explored the performance of one-slope solar still with sandbags filled with "El Qued sand." The results showed that modified solar still had an improvement of 34.57% in the cumulative yield and 34.83 % in the daily efficiency relative to the yield of solar still. According to Darbari and Rashidi [16], jute cloth as wick absorbers floating in basins improved the distillate output of solar still. With semicircular-shaped tooth absorbers, modified solar stills produced 65% more daily accumulation than SSS.

In the context of increasing yield of solar still by providing more wet surface area, Nagarajan et al. [17] explored the impact of adding baffles over absorber plates on the output of solar still. It has been reported that the adding baffles had an enhancement in yield by 58.82% when compared with solar still without baffles. Kumar et al. [18] tested the usefulness of finned basin and HSM on the yield of solar still. It has been found that efficiency is augmented by 64% and 95% with fins only and fins with HSM, respectively. Kabeel et al. [19] conducted experiments in a PSS with hollow circular fins and PCM and noticed that the yield is increased by 43% using hollow fins and 101.5% with both fins and PCM that that of SSS. Suraparaju and Natarajan [20] observed the effect of integrating solid staggered pin fins absorber inserted into PCM bed single-slope SS. The researchers found that the daily accumulation and thermal efficiency of modified solar still is augmented by 24.30% and 98.36% at a water depth of 2 cm, respectively, when compared to its smooth counterpart. Further, Suraparaju et al. [21] analyzed the influence of hollow fins and PCM on the accumulation of fresh water. The authors found that there is an increase in freshwater output of solar still by 15.7% for hollow fins only and 52.4% for hollow fins with PCM, respectively, relative to conventional SS. Recently, an in-depth review of solar stills integrated with sensible HSM has been presented by Khatod et al. [22] to find the optimum sensible HSM for a particular design of solar still. In the case of one-slope one-basin solar still, the CuO-coated absorber plate provided the highest yield (about 53% improvement over SSS). Whereas Kanchey marbles produce the maximum yield of 4,094 L/m^2/day for a two-slope one-basin SS. Recently, Sharma et al. [23] assessed the influence of copper fins positioned horizontally inside PSS on the overall freshwater accumulation. The authors reported that the copper fins increased the cumulative yield of PSS by 60%.

Based on the above studies, phase change materials used in solar stills have a remarkable influence on freshwater production. The use of PCMs inside copper fins in pyramid solar stills (PSS) has not been adequately studied. Therefore, this chapter investigates the productivity, thermal efficiency and payback period of PSS when copper fins of different heights and fins filled with PCM are mounted on the absorber plate.

8.2 EXPERIMENTAL FACILITY AND INSTRUMENTATION

Experimental work has been conducted in a PSS (see Figure 8.1) located on the rooftop of an independent building in Faridabad, Haryana (28.4035° N and 77.2894° E). The PSS is made of strong wood, which also act as an insulation material for retarding the heat loss to the surrounding, and its inner structure is overlaid with a galvanized iron (GI) sheet of 2-mm thickness that protects one-to-one interaction between wood and brine. Black paint is then applied to the GI sheet to enhance the heat absorption

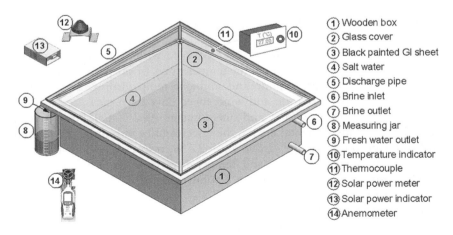

Figure 8.1 PSS with condensing cover and attached instruments.

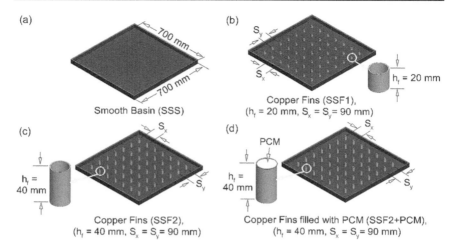

Figure 8.2 Configurations considered for improving evaporation rate.

and thereby accelerating the evaporation process. The PSS absorber basin had dimensions of 700 mm by 700 mm by 200 mm (length by breadth by height) and had an absorbing area of 0.49 m². Moreover, the solar still is covered with a pyramid-shaped condensing cover made of 3-mm glass, attached at an angle equal to the latitude of the testing side (i.e., 28.5") to ensure the best performance. The pyramid shape of condensation glass covers provides more surface space for vapour to condense on. A trough over the basin, which is connected by a flexible hose to the storage collecting jar, collects the distillate after the "evaporation-condensation process."

PSS absorber basin is adorned with copper fins of 20- and 40-mm height and PCM-filled copper tubes to augment the evaporation of salty water (see Figure 8.2). By strategically placing copper fins, freshwater yield is expected to be dramatically increased.

"K-type thermocouples" (0–1250°C) coupled with "digital temperature indicator" have been used to note the temperatures of salty water, absorber wall, inner glass cover, vapour and ambient temperatures. The experimental day's global radiation and wind speed have been measured using a Pyranometer (0–1800 W/m²) and anemometer (0–25 m/s), respectively. Considering all instrumental errors, the uncertainties in hourly and overall freshwater productivity are ~1.5% and ~2.5%, respectively. Furthermore, the errors in solar insolation and temperature readings are 2.88% and 0.58%, respectively. This results in a total error of 3.2% in the thermal efficiency of the PSS.

From April 10 to April 20, 2021, experiments have been conducted on SSS, SSF1, SSF2 and SSF2+PCM, and readings have been recorded every half an hour in order to determine how copper fins and PCM affect PSS

Thermal Efficiency Enhancement of Solar Still   145

productivity by augmenting salty water evaporation rates. There is salty water in the basin of PSS up to a depth of 30 mm.

The glass cover of the PSS allows solar energy to pass through and heat to be absorbed by the water as the sun rises. During the heating process, water vapour is released when enough energy is absorbed. During the condensation process, water vapour condenses on the glass surface due to temperature differences. The condensate is collected by sliding it over the inclined glass in a collecting trough. The use of solar stills is able to purify water for household purposes through evaporation and condensation.

8.3 THERMAL AND ECONOMIC ANALYSIS

The overall efficiency of the considered PSS in passive modes of operation is estimated by [9,19]:

$$\eta_{th} = \left[\frac{\sum m_w L}{A_s \int I(t)\,dt} \right] \times 100 \tag{8.1}$$

where L is the latent heat of vaporization (J/kg) and can be evaluated by using the following equation:

$$L = 2.4935 \times 10^6 [1 - (9.477 \times 10^{-4})T + (1.313 \times 10^{-7})T^2$$
$$- (4.497 \times 10^{-9})T^3] \text{ for } T < 70°C \tag{8.2}$$

Generally speaking, the issue of economic and financial feasibility is of prime concern for developing any solar thermal systems (STSs). That's why; the worth of PSS can be ultimately judged on the basis of its economy. The cost of freshwater production from brine using a passive solar still depends mainly on initial investment, rate of interest, useful lifecycle, average yearly freshwater productivity, and the running and maintenance cost [24]. Intentionally, the land cost to setup the PSS, and the yearly cost of feed water are overlooked. The cost breakdown of each component of PSS for all the considered cases, listed in Figure 8.2, is shown in Table 8.1. The parameters and equations considered for the economic analysis are provided henceforth [25].

If C_s be the initial capital cost invested in the fabrication of PSS with the rate of interest, i, per year ($i = 8\%$) and n, the useful lifecycle of a solar still in years (here, $n = 15$ years), Then, the capital recovery factor (CRF) and yearly fixed cost (C_{YFC}) can be calculated as:

$$CRF = \frac{i(1 + i)^n}{(1 + i)^n - 1} \tag{8.3}$$

146 Thermal Energy Systems

Table 8.1 Cost breakdown of PSS components

Component of PSS	SSS	SSF1	SSF2	SSF2+PCM
Glass cover	2,000	2,000	2,000	2,000
GI Sheet	2,000	2,000	2,000	2,000
Wooden frame	1,500	1,500	1,500	1,500
Collecting jar	370	370	370	370
Metal frame	1,500	1,500	1,500	1,500
Paint	470	470	470	470
PVC pipe	100	100	100	100
PCM material	–	–	–	300
Copper tubes	–	480	680	680
Total	**7,940**	**8,420**	**8,620**	**8,920**

$$C_{YFC} = C_s \times CRF \tag{8.4}$$

Yearly operational and maintenance cost (C_{YOMC}), including the amount spent in regular cleaning, filling of feed water, freshwater collection and removal of scaling because of salt deposition inside the PSS, is considered as 15% of C_{YFC} and presented as:

$$C_{YOMC} = 0.15\,(C_{YFC}) \tag{8.5}$$

The sinking fund factor (SFF) is calculated as:

$$SFF = \frac{i}{(1 + i)^n - 1} \tag{8.6}$$

Thus, the yearly salvage cost is calculated as:

$$C_{YSV} = S \times SFF \tag{8.7}$$

where S is the salvage cost, which is considered as 20% of C_s.
   Total yearly cost (C_{TYC}) is calculated by

$$C_{TYC} = C_{YFC} + C_{YOMC} - C_{YSV} \tag{8.8}$$

where C_{YFC}, C_{YOMC}, and C_{YSV} are the yearly fixed, operation and maintenance, and salvage costs, respectively.
   Average yearly freshwater yield in L/m^2 $\left(M_y\right)$ can be evaluated as:

$$M_y = m_w \times N_d \tag{8.9}$$

where m_d refers to the mean daily freshwater productivity per unit area and N_d refers to the number of clear days, that is, 330 days, for Kanchikacherla, India.

"Cost per litre" of fresh water produced (C_{CPL}) is estimated as:

$$C_{CPL} = \frac{\text{Total Yearly Cost}(C_{TYC})}{\text{Total Yearly Fresh Water Productivity}(M_y)} \quad (8.10)$$

Yearly market cost of distillate per year (C_{ymd}) = M_y × Market cost of water per kilogram

Net yearly earnings (C_{nye}) is calculated as:

$$C_{nye} = C_{ymd} - C_{YOMC} \quad (8.11)$$

Therefore, the payback period (N_P), in days, of the PSS is given by:

$$N_P = \frac{C_s}{C_{nye}} \times 365 \quad (8.12)$$

8.4 RESULTS AND DISCUSSION

For the days of conducting tests (April 10–20, 2021), climatological conditions characterized by "solar radiation intensity distribution" will be observed from 7:00 AM to 7:00 PM. Looking at Figure 8.3, the solar insolation rises gradually in the morning until it reaches its peak (911 W/m²) at

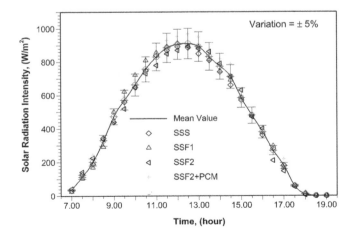

Figure 8.3 Variation in solar radiation distribution on all testing days.

around 12:30 AM, and then falls over time until reaching its minimum value at 6:00 PM due to sunset. In Figure 8.3, the experimental readings of solar radiation are compared, which demonstrate the effect of ambient conditions on the experimental results, thus indicating that the experimental results for various cases can be appropriately compared in order to figure out the impact of circular hollow and PCM-filled fins on the performance of PSS.

Figure 8.4 presents the variations of inner glass cover and basin water temperatures for SSS, SSF1, SSF2 and SSF2+PCM against time in order to understand the overall productivity variation for the considered cases. Both temperatures rise first, reaching their maximum nearby 1:30 PM and then decline thereafter in all cases. The highest temperature and solar irradiance

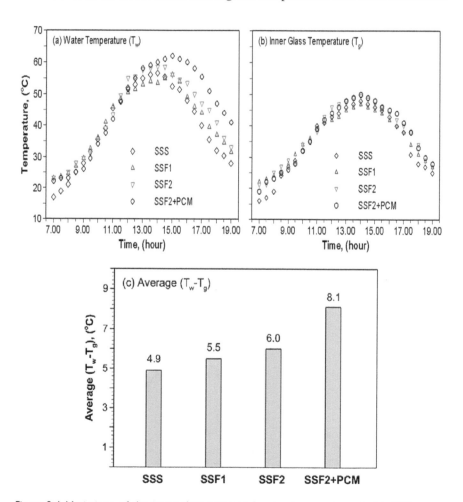

Figure 8.4 Variations of the inner glass cover and water temperatures inside PSS.

values are still separated by a time lag, which can be attributed to the time necessary for the solar still parts to warm.

The uppermost water temperatures of the SSS, SFF1, SFF2 and SSF2+PCM are logged as 55.5°C, 56.5°C, 58.0°C and 59.5°C, respectively (Figure 8.4a). Interestingly, the water temperature of SSF2+PCM is considerably higher than other cases, especially in evening hours, which can be associated with heat released by the PCM filled in copper fins to the water available inside the PSS. The inner glass cover temperatures of the SSS, SFF1, SFF2 and SSF2+PCM are logged as 48.0°C, 47.0°C, 49.5°C and 50.0°C, respectively (Figure 8.4b). Figure 8.4c indicates the average water and glass temperature difference, average $T_w - T_g$, which is of immense importance to foresee improvement in productivity. The higher difference in average $T_w - T_g$ signifies a higher rate of condensation over the glass cover. It is evident from Figure 8.4c that the maximum difference (8.1°C) in average $T_w - T_g$ of the PSS is observed for SSF2+PCM, followed by SSF2, SSF1 and SSS, respectively. The substantial values of average $(T_w - T_g)$ lead to augmentation in the rate of vapour condensation, and therefore enhancing the freshwater yield of SSF1, SSF2 and SSF2+PCM when compared to SSS.

Figure 8.5 illustrates how the cumulative production flux varies across all cases. Initially, the yields for all considered cases are in close proximity, but with progress in time, PSS with hollow vertical fins produces more fresh water because of wet surface area. However, from 3:00 PM onwards, the yield for SSF2+PCM is the highest which can be attributed to the heat released by the PCM and added to salty water in evening hours (as can be seen from higher water temperature for SSF2+PCM in Figure 8.4), leading to higher evaporation rate and subsequent enhancement in yield for SSF2+PCM than other cases.

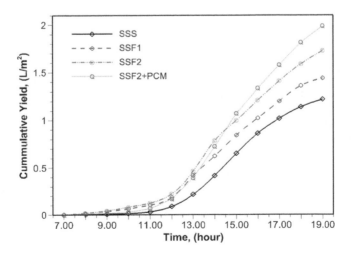

Figure 8.5 Variation of cumulative yield for studied cases.

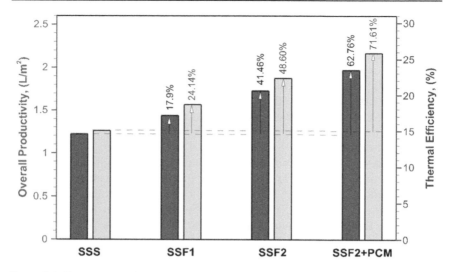

Figure 8.6 Comparison of productivity and thermal efficiency of PSS for explored cases.

As can be seen in Figure 8.6, SSF2+PCM offers the maximum yield (1.9857 L/m^2), followed by SSF2 (1.7258 L/m^2), SSF1 (1.4382 L/m^2) and SSS (1.22 L/m^2). For SSF2+PCM, SSF2 and SSF1, the PSS freshwater output is 62.76%, 41.46% and 17.9% more relative to SSS. The combined effect of thermal energy stored in the PCM and increased wet surface area results in a high heat transfer rate thereby leading to noticeably higher water temperatures in the evening (refer Figure 8.4) and, consecutively, a rise in evaporation rate that results in enhancement in freshwater output. Compared to SSF2, 15.06% growth in distillate yield is witnessed for the PCM-filled copper fins (SSF2+PCM).

The overall efficiency of PSS in all cases is assessed using equation (8.1) and shown as a bar diagram in Figure 8.6. The efficiencies of SSS, SSF1, SSF2 and SSF2+PCM are about 15.05%, 16.68%, 22.36% and 25.83%, respectively. The thermal efficiency of SSF2+PCM is about 71.61% higher than SSS. In conclusion, copper fins with and without PCM in the absorbent basin result in improved PSS performance. Energy efficiency and productivity have increased as a result of higher system temperatures.

Table 8.2 documents the economic viability of all considered cases based on economic analysis. The findings of the cost-effective analysis in this chapter in terms of CPL and payback period are shown in Figure 8.7. From Figure 8.7, it is evident that the CPL (Rs/L) is about 5, 4.5, 3.8 and 3.4 and the payback periods (days) are about 366, 329, 288 and 251 days for SSS, SSF1, SSF2 and SSF2+PCM, respectively.

As a result of the cost analysis, SSF2+PCM is found to have the lowest CPL, that is, about 32%, of distillate with the least payback period, that is, 31.42%, when compared to SSS. Altogether, the inclusion of fins and PCM in the absorbent basin is cost-effective and environmentally sustainable.

Table 8.2 Cost estimation of the PSS for studied cases

Comparison of economics	SSS	SSF1	SSF2	SSF2+PCM
Initial Investment (C_s), Rs	7,940	8,420	8,620	8,920
Yearly fixed cost (C_{YFC}), Rs	928	984	1,007	1,042
Yearly maintenance cost (C_{YOMC}), Rs	139	148	151	156
Yearly salvage cost (C_{YSV}), Rs	58	62	63	66
Annual cost, Rs	1,008	1,069	1,095	1,133
Annual productivity, L/m²	402.60	474.54	569.58	655.38
Cost per litre of distillate, Rs/L	5.0	4.5	3.8	3.4
Net yearly earning, Rs	7,913	9,343	11,241	12,951
Payback period (N_P), Days	366	329	280	251

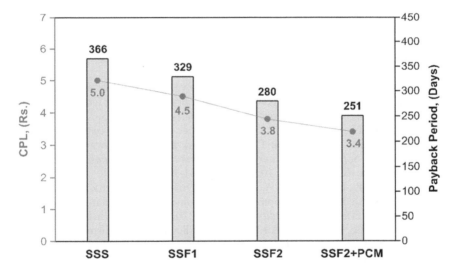

Figure 8.7 Results of economic analysis.

8.5 CONCLUSIONS

This study examines the effect of the hollow and PCM-filled copper fins over the basin on the productivity, thermal efficiency and economic aspects of PSS. Conclusions drawn from the present study are as follows:

- Use of hollow and PCM-filled copper fins in the basin increases the water temperature, up to 65.30% for SSF2+PCM as compared to SSS.
- The overall accumulations of SSS, SSF1, SSF2 and SSF2+PCM are about 1.22, 1.4382, 1.7258 and 1.9857 L/m², respectively.

152 Thermal Energy Systems

- An increase of 62.76% by SSF2+PCM, 41.46% by SSF2 and 17.90% by SSF1 in the overall yield is attained in comparison to SSS.
- Thermal efficiencies of SSF2+PCM, SSF2 and SSF1 are about 71.61%, 48.60% and 24.14% higher than SSS, respectively.
- CPL of fresh water is about 5, 4.5, 3.8 and 3.4 Rs/L, and the payback periods are about 366, 329, 280 and 251 days for SSS, SSF1, SSF2 and SSF2+PCM, respectively.

The above observations indicate that PCM-filled copper fins have the potential to desalinate salty water in an environmentally friendly manner.

REFERENCES

1. Sharma, N., Varun, Siddhartha.: Stochastic techniques used for optimization in solar systems: A review. *Renewable and Sustainable Energy Reviews*, 16(3), 1399–1411(2012).
2. Siddhartha, Sharma, N., Varun.: A particle swarm optimization algorithm for optimization of thermal performance of a smooth flat plate solar air heater. *Energy*, 38, 406–413 (2012).
3. Sunil, Varun, Sharma, N.: Experimental investigation of the performance of an indirect-mode natural convection solar dryer for drying fenugreek leaves. *Journal of Thermal Analysis and Calorimetry*, 118(1), 523–531 (2014).
4. Sharma, N., Choudhary, R.: Multi-objective performance optimization of a ribbed solar air heater. In: Tyagi, H., Chakraborty, P., Powar, S., Agarwal, A. (eds.) *Solar Energy. Energy, Environment, and Sustainability*. Springer, Singapore 77–93 (2020). 10.1007/978-981-15-0675-8_6
5. Singh, V.P., Jain, S., Karn, A., Kumar, A., Dwivedi, G., Meena, C.S., Dutt, N., Ghosh, A.: Recent developments and advancements in solar air heaters: A detailed review. *Sustainability*, 14, 12149 (2022). 10.3390/su141912149
6. Mohiuddin, S.A., Kaviti, A.K., Rao, T.S., Sikarwar, V.S.: Historic review and recent progress in internal design modification in solar stills. *Environmental Science and Pollution Research*, 1–54 (2022). 10.1007/s11356-022-19527-x
7. Singh, V.P., Jain, S., Karn, A., Kumar, A., Dwivedi, G., Meena, C.S., Cozzolino, R.: Mathematical modeling of efficiency evaluation of double-pass parallel flow solar air heater. Sustainability, 14, 10535 (2022). 10.3390/su141710535
8. Sharma, N., Noushad, S., Reddy, G.S.R.K., Ajit.: Productivity improvement of solar still using cemented blocks, International Conference on Advances in Energy Research (ICAER-2022), IIT Bambay, Maharashtra, July 7–9, 2022.
9. Natarajan, S.K., Suraparaju, S.K., Elavarasan, R.M., Pugazhendhi, R., Hossain, E.: An experimental study on eco-friendly and cost-effective natural materials for productivity enhancement of single slope solar still. *Environmental Science and Pollution Research*, 29(2), 1917–1936 (2022).
10. Panchal, H., Patel, D.K., Patel, P.: Theoretical and experimental performance analysis of sandstones and marble pieces as thermal energy storage materials inside solar stills. *International Journal of Ambient Energy*, 39(3), 221–229 (2018).

11. Sakthivel, T.G., Arjunan, T.V.: Thermodynamic performance comparison of single slope solar stills with and without cotton cloth energy storage medium. *Journal of Thermal Analysis and Calorimetry*, 137(1), 351–360 (2019).

12. Modi, K.V., Modi, J.G.: Performance of single-slope double-basin solar stills with small pile of wick materials. *Applied Thermal Engineering*, 149, 723–730 (2019).

13. Shehata, A.I., Kabeel, A.E., Dawood, M.M.K., Elharidi, A.M., Abd_Elsalam, A., Ramzy, K., Mehanna, A.: Enhancement of the productivity for single solar still with ultrasonic humidifier combined with evacuated solar collector: An experimental study. *Energy Conversion and Management*, 208, 112592 (2020).

14. Yarramsetty, N., Sharma, N., Narayana, M.L.: Experimental investigation of a pyramid type solar still with porous material: Productivity assessment. *World Journal of Engineering* (2021). 10.1108/WJE-02-2021-0096

15. Singh, V.P., Jain, S., Kumar, A. (2022). Establishment of correlations for the thermo-hydraulic parameters due to perforation in a multi-V rib roughened single pass solar air heater. *Experimental Heat Transfer*, 10.1080/08916152. 2022.2064940.

16. Darbari, B., Rashidi, S.: Performance analysis for single slope solar still enhanced with multi-shaped floating porous absorber. *Sustainable Energy Technologies and Assessments* 50, 101854 (2022).

17. Nagarajan, P.K., El-Agouz, S.A., DG, H.S., Edwin, M., Madhu, B., Sathyamurthy, R., Bharathwaaj, R.: Analysis of an inclined solar still with baffles for improving the yield of fresh water. *Process Safety and Environmental Protection*, 105, 326–337 (2017).

18. Kumar, S.T.R., Jegadheeswaran, S., Chandramohan, P.: Performance investigation on fin type solar still with paraffin wax as energy storage media. *Journal of Thermal Analysis and Calorimetry*, 136(1), 101–112 (2019).

19. Kabeel, A.E., El-Maghlany, W.M., Abdelgaied, M., Abdel-Aziz, M.M.: Performance enhancement of pyramid-shaped solar stills using hollow circular fins and phase change materials. *Journal of Energy Storage*, 31, 101610 (2020).

20. Suraparaju, S.K., Natarajan, S.K.: Experimental investigation of single-basin solar still using solid staggered fins inserted in paraffin wax PCM bed for enhancing productivity. *Environmental Science and Pollution Research*, 28(16), 20330–20343 (2021).

21. Suraparaju, S.K., Sampathkumar, A., Natarajan, S.K.: Experimental and economic analysis of energy storage-based single-slope solar still with hollow-finned absorber basin. *Heat Transfer*, 50(6), 5516–5537 (2021).

22. Khatod, K.J., Katekar, V.P., Deshmukh, S.S.: An evaluation for the optimal sensible heat storage material for maximizing solar still productivity: A state-of-the-art review. *Journal of Energy Storage*, 50, 104622 (2022).

23. Sharma, N., Noushad, S., Reddy, G.S.R. Kumar.: *Effect of Copper Fins on Fresh Water Productivity of Pyramid Solar Still.* pp. 83–91, Springer, Singapore (2022).

24. Ranjan, K.R., Kaushik, S.C.: Economic feasibility evaluation of solar distillation systems based on the equivalent cost of environmental degradation and high-grade energy savings. *International Journal of Low-Carbon Technologies*, 11(1), 8–15.

25. Tiwari, G.N.: Economic analysis of some solar energy systems. *Energy Conversion and Management*, 24(2), 131–135 (1984).

Chapter 9

Thermal and Electrical Management of a Solar PV/T System with and without PCM

Ankit Dev, Ravi Kumar, R. P. Saini, and Aditya Kumar

CONTENTS

9.1 Introduction 155
9.2 System Configuration and Methodology 157
9.3 Nanofluid Preparation 160
9.4 Methodology 160
9.5 Results and Discussion 161
9.6 Conclusion 166
Acknowledgements 167
References 167

9.1 INTRODUCTION

Primary conventional energy recourses (oil, natural gas and coal) play a major role to meet the energy demand of the world [1]. However, the consumption of these limited resources leads to the increment in CO_2 emission into the environment which causes grave consequences to the climate and increases the global temperature [2,3]. Alternatively, renewable energy sources have been successfully applied to various streams of fundamental sciences and engineering applications. The solar energy, biomass and wind energy are some of the examples of renewable energy resources and are abundantly available [4]. Photovoltaic (PV) systems for electricity generation have been proven of significant importance to the various sectors of electric energy applications for the sustainable development of society [5,6]. Thus far, PV panels have been applicable and used by the power sectors, agriculture sector, cottage industries, commercial and social services sectors [7]. The PV panels use around 10–15% of incident radiation to generate the electricity and the rest is absorbed by the panel causing the increment in the panels' temperature [8]. The electrical efficiency has shown a drastic deterioration with an increment in panel temperature. Electrical efficiency decreases by approximately 0.45% with the increase in every degree Celsius temperature [9]. To minimize this temperature increment,

DOI: 10.1201/9781003395768-9

cooling of the panel surface needs to be done from the backside. A pragmatic technique to anticipate overheating of PV modules is inserting a thermal energy absorber to disregard the exhaust heat. This joined framework is known as the photovoltaic thermal (PV/T) collector, which can be used to generate electricity as well as to use the removed heat [10]. This method of cooling the PV panel can be named active cooling as the fluid continuously flows through the thermal absorber with the help of a pump.

One more strategy to hold down the temperature of solar cells and to enhance the energy harvesting of the solar panel is the utilization of phase change materials (PCM). To utilize the PCM in cooling of PV panels, a layer of the PCM is placed at the back of the panel inside a metal container. This PCM layer will absorb the extra heat and store it in the form of latent heat. As this cooling method does not require any external power source can be named passive cooling of PV panels. A PCM is a material that has a high value of latent heat as a result of which it can absorb and release a lot of energy at a specified temperature [11] and its stored heat can be utilized for several heating applications [12]. The stored thermal energy due to latent heat has a higher energy density when compared to sensible energy [13]. For PV/T, a phase change material numerical model has been developed by Kibria et al. [14]. They observed that phase change materials are a very effective way to decrease the surface temperature of the solar cell. Apart from this, they noticed that the thermal performance of solar cells increased up to 5%. Recently, Saxena et al. [15] designed and investigate the performance of the PV system coupled with the heat sink structure with PCM in an innovative way. The reduction in the averaged surface temperature is noted at approximately 78.7% in the first configuration, 25.7% in the second and 36.8% in the third configuration. Pichandi et al. [16] reported an improvement in the performance of the PV module with the integration of inorganic PCM. They reported an instantaneous temperature reduction of 7°C and 1.21% increment in daily average efficiency. Marudaipillai et al. [17] investigated the thermal management and performance chrematistics of the PV panel with PCM. They have reported the enhancement of 3.667% in the panel's overall efficiency.

The use of advanced fluids such as nanofluids are one of a kind of advanced fluids having improved thermophysical properties show some promising results regarding the overall efficiency of the solar PV/T system. The colloidal suspension of the higher thermal conductivity solid nanoparticles into the conventional base fluids is known as nanofluids. Recently, Lee et al. [18] performed an efficiency analysis of the PV/T panels with CuO/water and Al_2O_3/water nanofluids as a working fluid and reported an increment of around 20% in the thermal and 0.07% in the electrical efficiency of the panel. Similarly, Waeil et al. [19] reported the overall efficiency of PV/T panels using SiC/water nanofluid around 88.9%, which is way higher than PV panel systems. Shamani et al. [10] have used the SiO_2, TiO_2 and SiC nanofluids and reported the highest thermal and electrical efficiency with SiC around 81.73% and 13.52%, respectively. Mustafa et al. [20] investigated the PV system's performance using TiO_2–water nanofluid and reported that panels show enhancement of 0.55% and 1.17% compared to the water and air flow channels, respectively. These studies encouraged to conduct more investigations on these aspects.

Several studies have been performed by researchers to find the effect of either active cooling or PCM cooling of the solar PV panel. Only a few investigations are available in the abundance of literature which are focused on combined active and PCM cooling of solar PV panels. In order to understand and accurately predict the PV/T characteristics trend more experimental investigations are required with various parameters. The primary objective of this investigation is to evaluate the performance of the PV/T system under the active cooling with and without PCM using different heat transfer fluids. Second, the objectives are expanded to characterize the used nanofluids and investigate the effect of their thermal conductivity on the performance of PV/T systems in a sustainable manner. This chapter includes the analysis of experimental results of the performance of hybrid PV/T at four different solar irradiations using five different heat transfer fluids. The experimental results of different PV/T system configurations with and without PCM are compared with the conventional PV panel and the improvement in the performance of PV panel has been noted.

9.2 SYSTEM CONFIGURATION AND METHODOLOGY

In order to achieve the proposed objectives, an experimental set-up of a PV/T system with a PCM container attached to the back of it has been designed, fabricated and installed at Renewable Energy R&D Laboratory. Complete specifications of the PV/T collector are presented in Table 9.1. The schematic

Table 9.1 Specifications of the PV/T collector

Items/Properties	Specifications
Type of cell	Monocrystalline
Number of cells	36
PV panel area (m^2)	0.38
PV panel maximum capacity (W)	35
Open circuit voltage (V)	22
Short circuit current (A)	2.5
Filled factor	0.636
Optimum power voltage (V)	22.6
Optimum operating current (A)	1.69
Absorber type	Copper tube
Thermal absorber area (m^2)	0.35

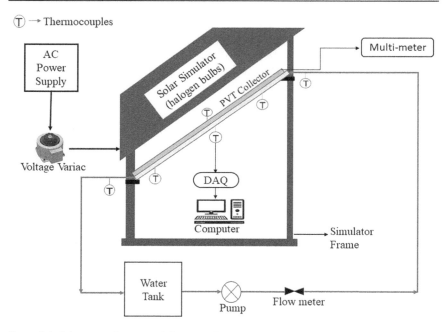

Figure 9.1 Schematic diagram of the complete experimental set-up.

diagram and photograph of the experimental set-up are shown in Figures 9.1 and 9.2, respectively. The experimental test set-up comprised of different components, i.e., solar simulator, PV/T-PCM, fluid storage tank, fluid circulation pump collector and a flow meter. This complete experimental set-up was connected with various instruments, i.e., digital solarimeter, voltage variac, multimeter and a data logger with a computer. The solar simulator used for the study was fabricated in-house using halogen bulbs each having a power rating of 200 W. The solar simulator was also connected to a voltage variac of range 0–260 V to vary the intensity of solar irradiation. The system was tested at four different values of solar irradiations, i.e., 450, 600, 850 and 1050 W/m^2. The experimental test section was fabricated by combining one PV panel with a PCM storage container at the backside of the panel. One thermal absorber made up of copper pipes was also fitted inside the PCM container. The container was closed from the bottom by using a Perspex sheet and made leak-proof by using silicon sealant. The complete PV/T-PCM collector is shown in Figure 9.3. A multimeter was connected to the PV panel to record the electric output from it. Finally, a fluid storage tank of 5 Litres capacity was connected to the thermal absorber through a small fluid flow pump having a power rating of 35 W to circulate the heat transfer fluid continuously through the absorber. A Rotameter was used as a flow meter having a range of 1–20 LPM. The heat transfer fluids used for the active cooling are water, Al_2O_3–water nanofluid and TiO_2–water nanofluid. Total

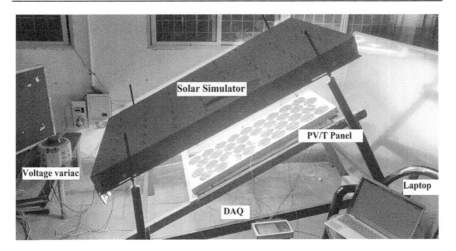

Figure 9.2 Photographic view of the experimental set-up.

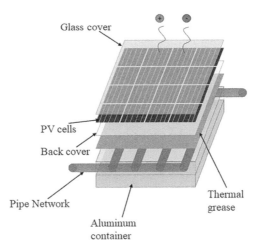

Figure 9.3 Schematic of the experimental test section.

four nanofluids have been prepared by using two different volume fractions (0.05% v/v and 0.1% v/v) of both Al_2O_3 and TiO_2 nanoparticles. The aluminium container that is attached to the backside of the panel containing PCM is filled with the phase change material (PCM) extracted the excess heat from the PV back surface and stores it in the form of latent heat and transfers it to the heat transfer fluid flowing inside the thermal absorber. An organics type paraffin wax is used in the study as a PCM reason being it is inexpensive, stable, non-toxic, environmentally harmless and abundantly available for different ranges of temperature The thermal and physical properties of the PCM are presented in Table 9.2.

Table 9.2 Thermophysical properties of the PCM material used in the study

Property	Value
Density solid	880 kg/m^3
Density liquid	760 kg/m^3
Melting temperature	50°C
Latent heat capacity	168 kJ/kg
Specific heat	2 kJ/kg K
Thermal conductivity	0.2 W/m K

9.3 NANOFLUID PREPARATION

The two-step approach was used for synthesizing of nanofluids in this chapter, as shown in Figure 9.4. The first step involves the preparation of nanoparticles either by chemical or physical method. The second step involves the uniform dispersion of those nanoparticles into the conventional base fluid by ultrasonication. Al_2O_3 and TiO_2 nanoparticles were used in the base fluid water to prepare the nanofluids. Both the nanofluids are prepared with two nanoparticle concentrations of 0.1% v/v and 0.5% v/v.

9.4 METHODOLOGY

The methodology used in this chapter to achieve the proposed objectives is explained in the following steps:

i. The range of design parameters and operating parameters are selected based on the literature review.
ii. Selection and characterization of suitable PCM have been done for the purpose of passive cooling of PV panels.

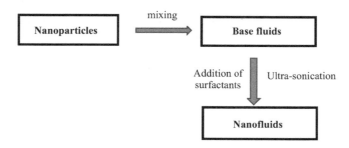

Figure 9.4 Schematic of the two-step method to prepare nanofluids in this study.

iii. Preparation and characterization of different types of stable nano-fluids have been done to study the active cooling of PV panels.
iv. Experimental setup is designed and fabricated to covert the conventional PV panel into PV/T-PCM system to perform the desired cooling.
v. Experimentations are carried out for each heat transfer fluid at all four solar irradiations with and without using PCM.
vi. The obtained data from experimentations are reduced in terms of temperature and electrical efficiency.
vii. The effects of system parameters and operating parameters like solar irradiation value, thermal conductivity of heat transfer fluid on the performance of solar PV panels are presented, analyzed and discussed.

The data of the investigation is analyzed by the following equations:
The maximum power generated from the PV panel is

$$P_{max} = V_{max} \times I_{max} \tag{9.1}$$

where I_{max} is the maximum operating current and V_{max} is the maximum power voltage.

Expression for electrical efficiency of the PV panel is represented as follows:

$$\eta_e = \frac{P_{max}}{I_s\, A_{panel}} \tag{9.2}$$

where I_s is the value of solar irradiation and A_{panel} is the PV panel surface area.

9.5 RESULTS AND DISCUSSION

The data were retrieved using an experimental test rig by following the methodology discussed earlier for the defined range of system and operating parameters. The main objective of this study was to evaluate and compare the performance of different PV/T systems under different cooling fluids with and without using PCM at four different solar irradiations. The data of PV/T system under each selected parameter are presented in terms of surface temperature, electrical efficiency and thermal efficiency. Based on the results presented a more efficient PV/T system can be selected in terms of cooling fluids along with suitable PCM.

Figure 9.5 presents the variation of the surface temperature of PV panels during water cooling with and without PCM at all the selected solar irradiations. These variations are compared with PV panels that is not provided with any cooling approach. The peak surface temperature got reduced

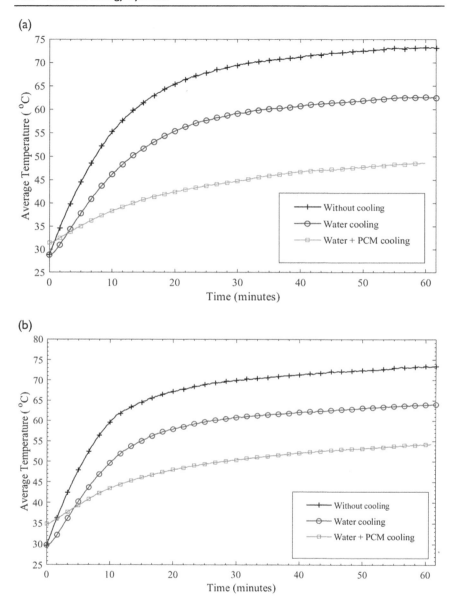

Figure 9.5 Variation of the surface temperature of the PV panel in case of water cooling with and without PCM when compared to the PV panel with no cooling at solar irradiations (a) 450 W/m², (b) 600 W/m², (c) 850 W/m² and (d) 1050 W/m².

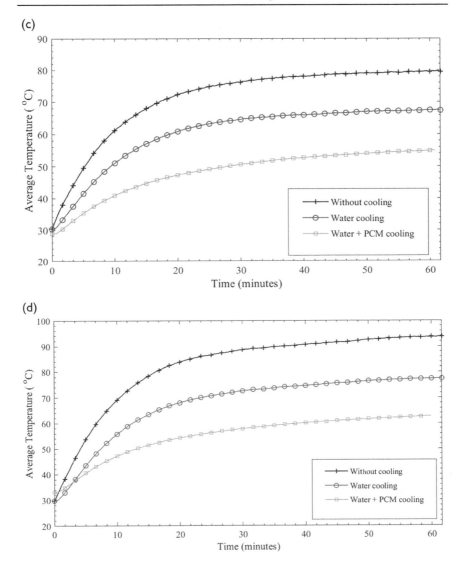

Figure 9.5 (*continued*)

by 16–18% in the case of only water cooling and 22–25% more reduction was reported in case of water cooling with PCM from 450 to 1050 W/m².

Figure 9.6 shows the electrical efficiency variation corresponding to each solar irradiation during no cooling, water cooling and water cooling with

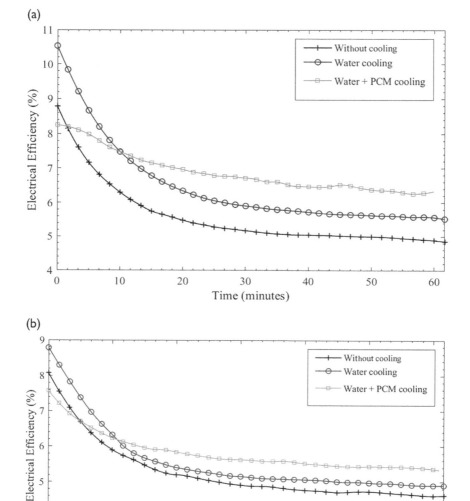

Figure 9.6 Variation of electrical efficiency of the PV panel in case of different cooling methods when compared to the PV system without cooling at solar irradiations (a) 450 W/m², (b) 600 W/m², (c) 850 W/m² and (d) 1050 W/m².

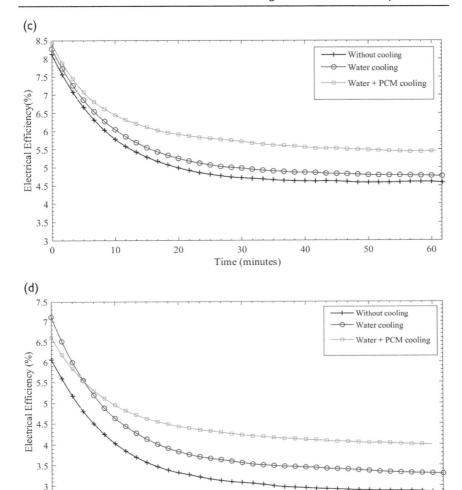

Figure 9.6 (continued)

PCM. As the surface temperature was increased, the efficiency reduces exponentially. During the cooling of PV panel, a significant improvement was reported in the efficiency due to temperature reduction of the panel surface. In case of water cooling the electrical efficiency at the time of peak surface temperature was improved by 15–21% from 450 to 1050 W/m². The water cooling with PCM further enhanced the electrical efficiency by 17–23% more as compared to only water cooling.

Figure 9.7 represents the effect of thermal conductivity of various heat transfer fluids on the electrical performance of a PV panel. Al_2O_3–water

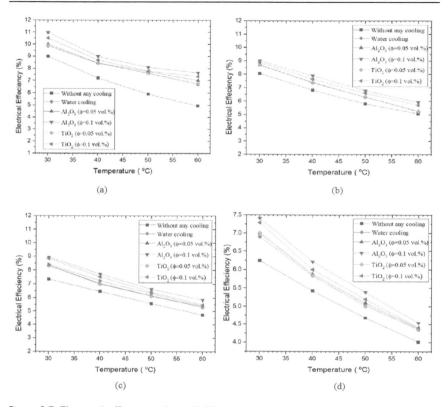

Figure 9.7 Electrical efficiency of the PV/T system in case of cooling with different heat transfer fluids when compared to the PV system without cooling at solar irradiations (a) 450 W/m^2, (b) 600 W/m^2, (c) 850 W/m^2 and (d) 1050 W/m^2.

nanofluids with 0.1% v/v has the highest thermal conductivity among all the fluids used for the study. The effect of the highest thermal conductivity can also be seen in Figure 9.7 also as the it produced highest electrical efficiency from the PV panel followed by the heat transfer fluids having thermal conductivity in decreasing order.

9.6 CONCLUSION

The above study was a performance evaluation test of a PV panel during different cooling methods at different solar irradiations. It was found that the combined active and passive cooling was the best when it comes to the reduction of the surface temperature of PV panel. The electrical efficiency of the panel was also improved with each cooling approach and found highest in case of combined cooling. The active cooling of the panel was further optimized by using different heat transfer fluids having different thermal conductivity and the fluids with the highest thermal conductivity produced maximum electrical output from the PV/T system. However, further research is required by using hybrid nanofluid and the using of various advanced PCM so that the solar PV/T system could be developed and applicable sustainably around the globe.

ACKNOWLEDGEMENTS

The authors are thankful to the Indian Institute of Technology Roorkee, India, for providing the necessary facilities and other backings for the research in every possible way. The authors have extended thanks to Dr Sudhakar Subudhi's group for providing the necessary support and equipment for research related to nanofluids.

REFERENCES

1. H. Ritchie, M. Roser, and P. Rosado, "Energy Mix," 2020. https://ourworldindata.org/energy#citation.
2. G. Fekadu and S. Subudhi, "Renewable energy for liquid desiccants air conditioning system: A review," *Renewable and Sustainable Energy Reviews*, vol. 93, pp. 364–379, 2018, 10.1016/j.rser.2018.05.016.
3. International Energy Agency Report: Global Energy Review, O_2 Emissions in 2021 Global emissions rebound sharply to highest ever level, 2021, 1–14.
4. A. Kumar, K. Kumar, N. Kaushik, S. Sharma, and S. Mishra, "Renewable energy in India: Current status and future potentials," *Renewable and Sustainable Energy Reviews*, vol. 14, no. 8, pp. 2434–2442, 2010, 10.1016/j.rser.2010.04.003.
5. A. H. A. Al-Waeli, K. Sopian, H. A. Kazem, and M. T. Chaichan, "Photovoltaic/thermal (PV/T) systems: Status and future prospects," *Renewable and Sustainable Energy Reviews*, vol. 77, pp. 109–130, 2017.
6. N. K. Malik, J. Singh, R. Kumar, and N. Rathi, "A review on solar PV cell," *International Journal of Innovative Technology Exploring Engineering*, vol. 3, pp. 116–119, 2013.
7. C. J. Smith, P. M. Forster, and R. Crook, "Global analysis of photovoltaic energy output enhanced by phase change material cooling," *Applied Energy*, vol. 126, pp. 21–28, 2014.
8. J. P. Ram, H. Manghani, D. S. Pillai, T. S. Babu, M. Miyatake, and N. Rajasekar, "Analysis on solar PV emulators: A review," *Renewable and Sustainable Energy Reviews*, vol. 81, pp. 149–160, 2018.
9. E. Skoplaki and J. A. Palyvos, "Operating temperature of photovoltaic modules: A survey of pertinent correlations," *Renewable Energy*, vol. 34, no. 1, pp. 23–29, 2009, 10.1016/j.renene.2008.04.009.

10. A. N. Al-Shamani, K. Sopian, S. Mat, H. A. Hasan, A. M. Abed, and M. H. Ruslan, "Experimental studies of rectangular tube absorber photovoltaic thermal collector with various types of nanofluids under the tropical climate conditions," *Energy Conversion and Management*, vol. 124, pp. 528–542, 2016, 10.1016/j.enconman.2016.07.052.

11. U. Stritih, "Increasing the efficiency of PV panel with the use of PCM," *Renewable Energy*, vol. 97, pp. 671–679, 2016.

12. S. S. Chandel and T. Agarwal, "Review of cooling techniques using phase change materials for enhancing efficiency of photovoltaic power systems," *Renewable and Sustainable Energy Reviews*, vol. 73, pp. 1342–1351, 2017.

13. S. H. Tasnim, R. Hossain, S. Mahmud, and A. Dutta, "Convection effect on the melting process of nano-PCM inside porous enclosure," *International Journal of Heat and Mass Transfer*, vol. 85, pp. 206–220, 2015.

14. M. A. Kibria, R. Saidur, F. A. Al-Sulaiman, and M. M. A. Aziz, "Development of a thermal model for a hybrid photovoltaic module and phase change materials storage integrated in buildings," *Solar Energy*, vol. 124, pp. 114–123, 2016.

15. A. Saxena, N. Agarwal, and B. Norton, "Design and performance characteristics of an innovative heat sink structure with phase change material for cooling of photovoltaic system," *Energy Sources, Part A: Recovery, Utilization and Environmental Effects*, pp. 1–25, 2021, 10.1080/15567036.2021.1968545.

16. R. Pichandi, K. Murugavel Kulandaivelu, K. Alagar, H. K. Dhevaguru, and S. Ganesamoorthy, "Performance enhancement of photovoltaic module by integrating eutectic inorganic phase change material," *Energy Sources, Part A: Recovery, Utilization and Environmental Effects*, pp. 1–18, 2020, 10.1080/15567036.2020.1817185.

17. S. K. Marudaipillai, B. Karuppudayar Ramaraj, R. K. Kottala, and M. Lakshmanan, "Experimental study on thermal management and performance improvement of solar PV panel cooling using form stable phase change material," *Energy Sources, Part A: Recovery, Utilization and Environmental Effects*, pp. 1–18, 2020, 10.1080/15567036.2020.1806409.

18. J. H. Lee, S. G. Hwang, and G. H. Lee, "Efficiency improvement of a photovoltaic thermal (PVT) system using nanofluids," *Energies (Basel)*, vol. 12, no. 16, p. 3063, Aug. 2019, 10.3390/en12163063.

19. A. H. A. Al-Waeli, K. Sopian, M. T. Chaichan, H. A. Kazem, H. A. Hasan, and A. N. Al-Shamani, "An experimental investigation of SiC nanofluid as a base-fluid for a photovoltaic thermal PV/T system," *Energy Conversion and Management*, vol. 142, pp. 547–558, 2017.

20. U. Mustafa, I. A. Qeays, M. S. BinArif, S. M. Yahya, and S. bin Shahrin, "Efficiency improvement of the solar PV-system using nanofluid and developed inverter topology," *Energy Sources, Part A: Recovery, Utilization and Environmental Effects*, pp. 1–17, 2020, 10.1080/15567036.2020.1808119.

Chapter 10

Second Law Analysis of Desiccant Cooling-Based Thermal Systems

D.B. Jani

CONTENTS

10.1 Introduction 169
10.2 System Operation.. 171
10.3 Energy Analysis.. 174
10.4 Results and Discussion... 177
10.5 Conclusions.. 178
References 179

10.1 INTRODUCTION

For the last many years, global warming and fossil fuel-based pollution creates many problems for mankind all over the world. The use of coal and petroleum-based conventional fuels increases day by day hiking the petroleum prices exponentially globally. Cooling power required for the building consumes almost 40% of the total energy requirements of the building. So, the innovative cooling demands to reduce the overall energy consumption by increasing energy efficiency for the cooling of buildings. Thus, built environment requirement mostly provided by the use of VCR-based conventional cooling systems which demands the use of high-grade electricity for its operation. So, advanced cooling technology almost demanded ameliorated cooling efficiency, the use of waste and renewable energy sources [1–3]. The requirement demanded by the modern cooling can be met successfully by the use of desiccant-assisted innovative cooling systems as it can make use of waste industrial heat or coupled to renewable solar energy. It is a newer approach in thermal cooling systems to resolve both energy and environmental issues as compared to conventionally used vapour compression-based traditional cooling [4].

The desiccant-based thermal cooling systems may be classified further according to the type and arrangement of its component are open-cycle desiccant dehumidification and cooling that makes use of open-air for

DOI: 10.1201/9781003395768-10

169

adsorption in dehumidifier while closed-cycle desiccant dehumidification system uses room air recirculated for dehumidification and cooling operation by dehumidifier and cooling coil, respectively. According to the nature of desiccant materials, their availability in nature, desiccant-powered dehumidification and cooling systems can be mainly classified as solid and liquid desiccant systems. In the case of rotary solid-desiccant dehumidification systems, dry solid-desiccant materials such as silica gel or zeolite can be used in a rotating bed or impregnated into a honeycomb-structure matrix form a wheel is known as a rotary dehumidifier within the system [5–7]. In case of liquid desiccant dehumidifiers, systems are innovative dehumidification and cooling technology consisting of a liquid film contact surface, which is either a closed cooling coil or an in-house cooling tower, mostly wetted with liquid desiccant dehumidification materials like lithium chloride or calcium chloride. Desiccant-based desorption and sensible cooling systems can be successfully used in humid environments with low cooling demand but have limited application in low-humidity areas where desiccants need not require the air moisture content control to the desired level for indoor built thermal comfort. In the Indian sub-continent, many residential and in-house commercial projects will be planned on the concept of desiccant desorption and cooling innovative technology. Many research projects are still underway to utilize the application more commercially and economically viable.

A solid-desiccant integrated VCR-assisted hybrid-built cooling system is experimentally investigated in a typical transient climate. The key components of innovative hybrid cooling that makes use of a rotary regenerator wheel, sensible heat recovery wheel, conventional vapour-compression split air-conditioner unit, data acquisition system and instrumentations. The experimental data has been collected and analyzed. The obtained experimental results are discussed in detail in result and discussion.

The system can be operated in two different configuration recirculation mode (closed) as well as in ventilation mode (open). For the study of performance evaluation, the recirculation mode is considered for the present study. In recirculation configuration, room air is recirculated as the process air side is a closed configuration while the regeneration air is in open atmospheric condition. Mostly the working of system in recirculation dust and infectants were separated out and exhausted to open air along with waste reactivation air to outdoor environment.

Moreover, the desiccant-powered thermal cooling systems are eco-friendly and consume minimum high-grade electricity for its operation as compared to the conventional system. Current work shows experimental assessment of desiccant-powered thermal cooling systems. The basic concept of exergy calculation under various components has been applied for second law analysis to assess this innovative cooling system.

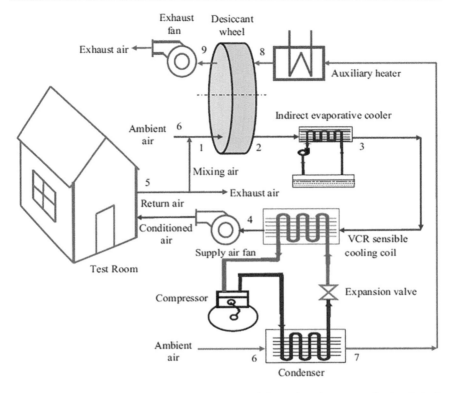

Figure 10.1 Schematic diagram of desiccant-powered thermal cooling system (recirculation).

10.2 SYSTEM OPERATION

The working of the system has been shown in Figures 10.1 and 10.2 schematically recirculation operating mode and ventilation mode set up, respectively. Recirculated room air initially passes through a dehumidifier for desorption and exiting in dry cooling and may pass further to a sensible heat wheel for sensibly cooling to achieve good efficiency in cooling coil for final condition before supplying to a room for air conditioning [8–11].

The regeneration air supply from open ambient conditions for initial passing through a sensible heat wheel for warming by utilizing heat rejected by process air pre-cooling in the same wheel. After passing through the heater coil, it heats up to the regeneration temperature that is necessary to drive off the moisture retained in the process, and air section is finally exhausted to the open atmosphere [12,13]. This modification is incorporated in the system in case of mild ambient. In the ventilation configuration

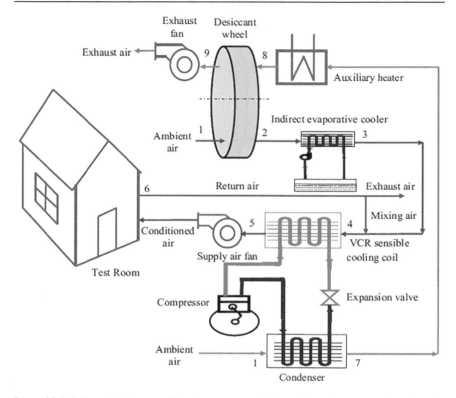

Figure 10.2 Schematic diagram of desiccant-powered thermal cooling system (ventilation).

instead of using room air ventilated air is used as process air and room air is used as reactivation air (Figure 10.2).

Conventional VCR systems may combine innovative desiccant cooling for the hot and humid transit climate (Figure 10.3). It requires downsized cooling unit as dehumidification (latent cooling load) has been carried out separately into a rotary desiccant dehumidifier unit so the cooling coil has to handle a sensible cooling load only. Upon reaching the criteria of temperature or humidity either of the system components dehumidifier or sensible cooling coil works as per on/off control provided through humidistat or thermostat monitoring unit installed at the test room [14–16].

The desiccant integrated rotary dehumidifier used in the above hybrid cooling system is one of the key components of the solid-desiccant VCR hybrid innovative cooling built environment system for effectively maintaining the indoor humidity of room recirculated process air. Other necessary components required are a wheel driving motor, a wheel disk and a drive belt. The technical specifications of the dehumidifier used in the present system are shown in Table 10.1.

Second Law Analysis of Desiccant Cooling-Based Thermal Systems 173

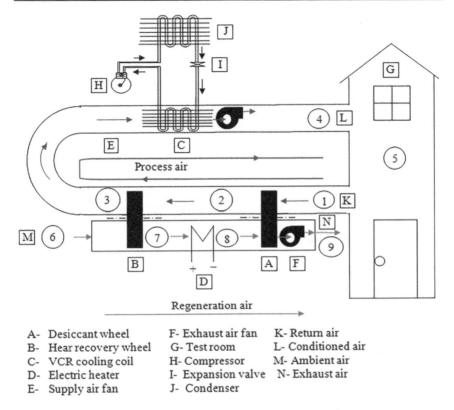

A- Desiccant wheel F- Exhaust air fan K- Return air
B- Hear recovery wheel G- Test room L- Conditioned air
C- VCR cooling coil H- Compressor M- Ambient air
D- Electric heater I- Expansion valve N- Exhaust air
E- Supply air fan J- Condenser

Figure 10.3 Schematic diagram of desiccant dehumidification integrated conventional vapour compression system.

Table 10.1 Technical specifications of desiccant wheel

Sr. No.	Parameter	Specification
1	Diameter	360 mm
2	Width	100 mm
3	Desiccant material	Synthesized metal silicate
4	Channel shape	Sinusoidal
5	Size of the mesh of rotor (thickness)	0.34 mm × 3.2 mm × 1.8 mm
6	Flow pattern inside the channels	Laminar flow
7	Cross area (P/R) ratio	3:1
8	Rotational speed	20 rph
9	Drive	230V/AC/50 Hz

10.3 ENERGY ANALYSIS

The performance evaluation of desiccant VCR-integrated hybrid dehumidification and cooling system depends on many factors such as changes in environmental conditions, air flow rates in process and reactivation sides respectively, pressure drop across dehumidifier, variations in human occupancies in the room and seasonal changes in the outdoor environment [17–21]. Performances indices are moisture removal rate, cooling capacity, coefficient of performance of the overall system, etc.

The important indicator for the performance evaluation of dehumidifiers is the moisture removal rate (MRR) and effectiveness (ε_{dw}) measured and calculated for considering process and reactivation air. The moisture removal rate or moisture transfer rate [22,23] is generally given as

$$MRR = \dot{m}_{pa}(\omega_1 - \omega_2)$$

where $\dot{m}_{pa}$ is the mass flow rate of process air at the desiccant wheel process air inlet, ω_1 and ω_2 are the humidity ratios of process air at the intake and exit of desiccant wheel, respectively. The effectiveness of the desiccant wheel is calculated by evaluating the ratio of the change in possible humidity ratio of the air to the maximum possible change in humidity ratio of the process air side of the dehumidifier.

The effectiveness of the rotary desiccant dehumidifier [24–26] has been denoted by the following equation:

$$\varepsilon_{dw} = \frac{\omega_1 - \omega_2}{\omega_1 - \omega_{2,ideal}}$$

The coefficient of performance of the system based on electrical supply energy to run the electrical motor energy input is calculated by knowing the ratio of the cooling capacity to the total electrical energy supply to the electric motor (E_{total}) is [27–29] mentioned by

$$COP = \frac{Q_{cc}}{E_{total}}$$

In the above equation, Q_{cc} is the cooling capacity of the sensible cooling coil [30–32] and is given as

$$Q_{cc} = \dot{m}_{pa}(h_1 - h_4)$$

where $\dot{m}_{pa}$ is the mass flow rate of process air at the desiccant wheel intake side. While total electric energy consumption by all above components can be summarized as E_{total} the total electrical power [33–35] is denoted as

$$E_{total} = E_{compressor} + E_{fan} + E_{heater} + E_{others}$$

where individual power for running the compressor and fan used in the system were $E_{compressor}$ and E_{fan}, respectively. Two fans for process and reactivation air circulations were used [36,37]. The notation for the electric power utility for running the reactivation heater is E_{heater}. Power consumption for the other auxiliary equipment is denoted as E_{others} [38–40].

Energy analysis described above states the energy balances of first law of thermodynamics but it does not show energy loss. This can be explained further in exergy analysis that includes second law analysis which explains both the quantity and quality of energy [41,42]. The rate of exergy can be given for the system as per the following expression:

$$\dot{E}_X = \dot{m}e_X = \dot{m}[(h - T_0s) - (h_0 - T_0s_0)]$$

where e is the specific exergy flow and state 0 represent the dead state, i.e., the outdoor environment.

The rate of exergy destruction found in the component can be expressed as

$$\dot{E}_{Xlost} = T_0\dot{S}_{gen}$$

where $\dot{S}_{gen}$ is the rate of entropy generation as per second law analysis in the given component of the system.

The rate of entropy generation found in the dehumidifier can be expressed as

$$\dot{S}_{gen,\ DW} = \dot{m}_p(s_2 - s_1) + \dot{m}_r(s_9 - s_8)$$

where $\dot{m}_p$ and $\dot{m}_r$ are rates of mass flow for the process or reactivation air flow in the dehumidifier. The rate of entropy generation in air-to-air heating sensible heating wheel can be given as

$$\dot{S}_{gen,\ HRW} = \dot{m}_p(s_3 - s_2) + \dot{m}_r(s_7 - s_6)$$

The rate of entropy generation under a cooling coil can be given as

$$\dot{S}_{gen,\ CC} = \dot{m}_p(s_4 - s_3) + \frac{\dot{Q}_{evop}}{T_6}$$

176 Thermal Energy Systems

where T_6 is the temperature of air in the open environment and the rate of heat removal in sensible cooling coil $\dot{Q}_{evop}$ can be expressed as

$$\dot{Q}_{evop} = \dot{m}_p(h_3 - h_4)$$

The rate of entropy generation for a thermal heater for reactivating the solid-desiccant material used in the dehumidifier can be given as

$$\dot{S}_{gen, HT} = \dot{m}_r(s_8 - s_7) - \frac{\dot{Q}_{reg}}{T_8}$$

where rate of regeneration for reactivation heat supply $\dot{Q}_{reg}$ can be expressed as

$$\dot{Q}_{reg} = \dot{m}_r(h_8 - h_7)$$

The exergy efficiency of the desiccant integrated cooling system [43–45] can be given as

$$\eta_{ex} = \frac{\dot{E}_{Xcool}}{\dot{E}_{Xheat} + \dot{E}_{Xcomp} + \dot{E}_{Xfan}}$$

where $\dot{E}_{cool}$ is the exergy difference obtained between the conditioned room and outdoor environment. This can be expressed as

$$\dot{E}_{Xcool} = \dot{m}_p[(h_1 - h_4) - T_6(s_1 - s_4)]$$

$\dot{E}_{heat}$ is the exergy input to supply reactivation heat supply in heater

$$\dot{E}_{Xheat} = \dot{m}_r[(h_8 - h_7) - T_6(s_8 - s_7)]$$

In the same way, the exergy efficiency of a dehumidifier, cooling coil, heater and sensible heat wheel can be given as

$$\eta_{DW} = \frac{\dot{m}_p[(h_2 - h_1) - T_6(s_2 - s_1)]}{\dot{m}_r[(h_8 - h_9) - T_6(s_8 - s_9)]}$$

Second Law Analysis of Desiccant Cooling-Based Thermal Systems   177

$$\eta_{HRW} = \frac{\dot{m}_r[(h_7 - h_6) - T_6(s_7 - s_6)]}{\dot{m}_p[(h_2 - h_3) - T_6(s_2 - s_3)]}$$

$$\eta_{CC} = 1 - \frac{\dot{E}_{XlostCC}}{\dot{E}_{X3}}$$

where $\dot{E}_3$ is the exergy rate for process air flow stream after passing through air to air heat recovery wheel at state point 3. The exergy efficiency for the thermal heater can be given by

$$\eta_{HT} = \frac{[(h_8 - h_7) - T_6(s_8 - s_7)]}{(h_8 - h_7) - T_6(h_8 - h_7)/T_8}$$

10.4 RESULTS AND DISCUSSION

The overall thermodynamic behaviour of the desiccant integrated thermal cooling system in terms of temperature, humidity as well as entropy obtained as inlet and exit components of the system is tabulated in Table 10.2.

Exergy efficiency and exergy loss in an individual component used in the hybrid thermal cooling system are provided in Table 10.3. It is seen from the table that the lowest exergy efficiency has been found in the case of cooling coil of approximately 8.9%. This is due to the presence of irreversibility during cooling. This may be due to the large temperature difference observed at the time of cooling between hot and cold fluid flow.

Table 10.2 Temperature, humidity and entropy at important state points in the system

State point	Temperature (°C)	Humidity ratio (g/kg)	Entropy (kJ/K kg da)
1	25.9	12.10	0.203
2	54.2	7.10	0.250
3	42.1	7.90	0.219
4	9.40	6.90	0.103
5	26.0	10.50	0.189
6	28.0	18.50	0.265
7	39.2	17.10	0.291
8	123.1	17.00	0.538
9	47.6	21.30	0.355

178 Thermal Energy Systems

Table 10.3 Exergy loss and exergy efficiency at the important component of the hybrid thermal cooling system

	Exergy destruction (kW)	Exergy efficiency (%)
Desiccant wheel	0.58	74.02
Heat recovery wheel	0.37	41.00
Heating system	0.57	53.16
Sensible cooling coil	0.14	8.97
System	1.68	11.45

Sensible heat recovery wheel can give exergy efficiency of approximately 41% due to irreversible mixing between hot and cold fluid in process air side and regeneration air side, respectively.

The highest exergy loss has been found in the desiccant wheel used in the thermal cooling system around 0.58 kW. The reason for the same is maximum irreversibility has been found due to slight mixing of cold process air and hot reaction air flow mixing in the desiccant wheel. By selecting the proper design of wheel to minimize the mixing and by properly optimizing wheel speed, these losses can reduce further. Losses in electric heating system which makes use of high-grade electricity are also found second most approximately 0.57 kW. The reason for the same is to use high-grade electricity for irreversible heating of the reactivation air to the desired high temperature demanded by the type of desiccant material used in the rotary dehumidifier.

Loss of exergy above both of the components can be reduced further by a selection of desiccant material in a rotary dehumidifier which possibly re-generated desiccant to the near existing ambient conditions by demanding lower heat for desorption will definitely decrease the irreversibility will contribute low exergy losses.

10.5 CONCLUSIONS

A second law analysis has been evaluated for the desiccant-assisted dehumidification and thermal cooling systems based on experimental data for various parameters like pressure drop, humidity ratios, temperatures, flow rates, etc. Exergy efficiency and exergy loss that occurred at different components used in the system have been calculated based on temperature, humidity, entropy, etc. From the obtained results, it has been found that the total exergy efficiency and exergy loss of the system are 11.45% and 1.68 kW, respectively. The desiccant wheel is the component having the most exergy loss (0.58 kW) among all the other components used in the system. The exergy loss of the system can be reduced further by selection of desiccant material used in a rotary dehumidifier which possibly regenerated desiccant to the near existing ambient conditions by demanding lower heat for desorption will definitely decrease overall irreversibility. Thus, by the study of second law analysis of desiccant-assisted thermal cooling systems the sites where high exergy losses can be identified and corrective action may be taken like improving design or better material selection to enhance exergy performance of the system.

REFERENCES

1. Bejan, A. (1980). Second law analysis in heat transfer. *Energy*, 5(8-9), 720–732.
2. San, J. Y., Worek, W. M., & Lavan, Z. (1987). Entropy generation in combined heat and mass transfer. *International Journal of Heat and Mass Transfer*, 30(7), 1359–1369.
3. Van den Bulck, E. S. A. K., Klein, S., & Mitchell, J. W. (1988). Second law analysis of solid desiccant rotary dehumidifiers. *The Journal of Solar Energy Engineering*, 110(1): 2–9. 10.1115/1.3268233.
4. Ahmed, C. K., Gandhidasan, P., Zubair, S. M., & Al-Farayedhi, A. A. (1998). Exergy analysis of a liquid-desiccant-based, hybrid air-conditioning system. *Energy*, 23(1), 51–59.
5. Camargo, J. R., Ebinuma, C. D., & Silveira, J. L. (2003). Thermoeconomic analysis of an evaporative desiccant air conditioning system. *Applied Thermal Engineering*, 23(12), 1537–1549.
6. Dincer, I., & Sahin, A. Z. (2004). A new model for thermodynamic analysis of a drying process. *International Journal of Heat and Mass Transfer*, 47(4), 645–652.
7. Rosen, M. A., & Dincer, I. (2004). Effect of varying dead-state properties on energy and exergy analyses of thermal systems. *International Journal of Thermal Sciences*, 43(2), 121–133.
8. Caliskan, H., Dincer, I., & Hepbasli, A. (2011). Exergetic and sustainability performance comparison of novel and conventional air cooling systems for building applications. *Energy and Buildings*, 43(6), 1461–1472.
9. Hürdoğan, E., Buyükalaca, O., Hepbasli, A., & Yılmaz, T. (2011). Exergetic modeling and experimental performance assessment of a novel desiccant cooling system. *Energy and Buildings*, 43(6), 1489–1498.
10. Koronaki, I. P., Rogdakis, E., & Kakatsiou, T. (2011). Thermodynamic analysis and neural network model of open cycle desiccant cooling systems with silica gel. *International Review of Mechanical Engineering*, 5(2), 298–304.
11. Enteria, N., Yoshino, H., Takaki, R., Yonekura, H., Satake, A., & Mochida, A. (2013). First and second law analyses of the developed solar-desiccant air-conditioning system (SDACS) operation during the summer day. *Energy and Buildings*, 60, 239–251.
12. Koronaki, I. P., Rogdakis, E., & Kakatsiou, T. (2013). Experimental assessment and thermodynamic analysis of a solar desiccant cooling system. *International Journal of Sustainable Energy*, 32(2), 121–136.

13. Rafique, M. M., Gandhidasan, P., Al-Hadhrami, L. M., & Rehman, S. (2016). Energy, exergy and anergy analysis of a solar desiccant cooling system. *Journal of Clean Energy Technologies*, 4(1), 78–83.
14. Jani, D. B., Mishra, M., & Sahoo, P. K. (2016). Solid desiccant air conditioning – A state of the art review. *Renewable and Sustainable Energy Reviews*, 60, 1451–1469.
15. Uçkan, İ., Yılmaz, T., Hürdoğan, E., & Büyükalaca, O. (2014). Exergy analysis of a novel configuration of desiccant based evaporative air conditioning system. *Energy Conversion and Management*, 84, 524–532.
16. Zendehboudi, A., (2016). Implementation of GA-LSSVM modelling approach for estimating the performance of solid desiccant wheels. *Energy Conversion and Management*, 127, 245–255.
17. Jani, D. B., Mishra, M., & Sahoo, P. K. (2017). Application of artificial neural network for predicting performance of solid desiccant cooling systems – A review. *Renewable and Sustainable Energy Reviews*, 80, 352–366.
18. Dadi, M., & Jani, D. B. (2019). TRNSYS simulation of an evacuated tube solar collector and parabolic trough solar collector for hot climate of Ahmedabad. Available at SSRN 3367102.
19. Jani, D. B. (2019). An overview on desiccant assisted evaporative cooling in hot and humid climates. *Algerian Journal of Engineering Technology*, 1, 1–7.
20. Peng, D., Zhou, J., & Luo, D. (2017). Exergy analysis of a liquid desiccant evaporative cooling system. *International Journal of Refrigeration*, 82, 495–508.
21. Jani, D. B., Mishra, M., & Sahoo, P. K. (2016). Exergy analysis of solid desiccant-vapour compression hybrid air conditioning system. *International Journal of Exergy*, 20(4), 517–535.
22. Guan, B., Liu, X., & Zhang, T. (2020). Analytical solutions for the optimal cooling and heating source temperatures in liquid desiccant air-conditioning system based on exergy analysis. *Energy*, 203, 117860.
23. Jani, D. B., Bhabhor, K., Dadi, M., Doshi, S., Jotaniya, P. V., Ravat, H., & Bhatt, K. (2020). A review on use of TRNSYS as simulation tool in performance prediction of desiccant cooling cycle. *Journal of Thermal Analysis and Calorimetry*, 140(5), 2011–2031.
24. Jani, D. B. (2019). An overview on recent development in desiccant materials. *International Journal of Advanced Materials Research*, 5 (2), 31–37.
25. Jani, D. B., Mishra, M., & Sahoo, P. K. (2018). Performance analysis of a solid desiccant assisted hybrid space cooling system using TRNSYS. *Journal of Building Engineering*, 19, 26–35.
26. Bhabhor, K. K., & Jani, D. B. (2019). Progressive development in solid desiccant cooling: A review. *International Journal of Ambient Energy*, 1–24.
27. Jani, D. B., Mishra, M., & Sahoo, P. K. (2017). A critical review on solid desiccant-based hybrid cooling systems. *International Journal of Air-conditioning and Refrigeration*, 25(03), 1730002.
28. Ge, F., & Wang, C. (2020). Exergy analysis of dehumidification systems: A comparison between the condensing dehumidification and the desiccant wheel dehumidification. *Energy Conversion and Management*, 224, 113343.
29. Jani, D. B., Mishra, M., & Sahoo, P. K. (2018). Investigations on effect of operational conditions on performance of solid desiccant based hybrid cooling system in hot and humid climate. *Thermal Science and Engineering Progress*, 7, 76–86.

30. Vyas, V., & Jani, D. B. (2016). An overview on application of solar thermal power generation. *International Journal of Engineering Research and Allied Sciences*, 1, 1–5.
31. Ahmad, T., & Chen, H. (2018). Short and medium-term forecasting of cooling and heating load demand in building environment with data-mining based approaches. *Energy and Buildings*, 166, 460–476.
32. Jani, D. B., Mishra, M., & Sahoo, P. K. (2016). Performance prediction of rotary solid desiccant dehumidifier in hybrid air-conditioning system using artificial neural network. *Applied Thermal Engineering*, 98, 1091–1103.
33. Bhabhor, K. K., & Jani, D. B. (2021). Performance analysis of desiccant dehumidifier with different channel geometry using CFD. *Journal of Building Engineering*, 103- 021.
34. Dadi, M., & Jani, D. B. (2019). Solar energy as a regeneration heat source in hybrid solid desiccant – vapor compression cooling system – A review. *Journal of Emerging Technologies and Innovative Research*, 6(5), 421–425.
35. Jani, D. B. (2022). TRNSYS simulation of desiccant powered evaporative cooling systems in hot and humid climate. *Journal of Mechanical Engineering*, 1(1), 1–6.
36. Zhang, Q., Liu, X., Zhang, T., & Xie, Y. (2020). Performance optimization of a heat pump driven liquid desiccant dehumidification system using exergy analysis. *Energy*, 204, 117891.
37. Jani, D. B., Mishra, M., & Sahoo, P. K. (2022). TRNSYS simulation of desiccant-assisted HVAC System. *Air conditioning and Refrigeration Journal*, 24(6), 28–33.
38. Jia, C. X., Dai, Y. J., Wu, J. Y., & Wang, R. Z. (2007). Use of compound desiccant to develop high performance desiccant cooling system. *International Journal of Refrigeration*, 30(2), 345–353.
39. Hua, L., & Wang, R. (2022). An exergy analysis and parameter optimization of solid desiccant heat pumps recovering the condensation heat for desiccant regeneration and heat transfer enhancement. *Energy*, 238, 121811.
40. Jani, D. B. (2021). A recent development in desiccant assisted HVAC system. *Air Conditioning and Refrigeration Journal*, 24(2), 26–52.
41. Jani, D. B. (2021). Use of artificial neural network in performance prediction of solid desiccant powered vapor compression air conditioning systems. *Instant Journal of Mechanical Engineering*, 3(3), 5–19.
42. Jani, D. B. (2019). Desiccant cooling as an alternative to traditional air conditioners in green cooling technology. *Instant Journal of Mechanical Engineering*, 1(1), 1–13.
43. Halliday, S. P., Beggs, C. B., & Sleigh, P. A. (2002). The use of solar desiccant cooling in the UK: A feasibility study. *Applied Thermal Engineering*, 22(12), 1327–1338.
44. Kanoğlu, M., Çarpınlıoğlu, M. Ö., & Yıldırım, M. (2004). Energy and exergy analyses of an experimental open-cycle desiccant cooling system. *Applied Thermal Engineering*, 24(5-6), 919–932.
45. Açıkkalp, E., Caliskan, H., Hong, H., Piao, H., & Seung, D. (2022). Extended exergy analysis of a photovoltaic-thermal (PVT) module based desiccant air cooling system for buildings. *Applied Energy*, 323, 119581.

LIST OF NOMENCLATURE

COP	Coefficient of performance
e_x	Sp. exergy of air stream (kJ/kg)
E	Electrical energy consumption (kW)
$\dot{E}_X$	Exergy rate (kW)
$\dot{E}_{lost}$	Rate of exergy loss (kW)
$\dot{E}_{comp}$	Rate of exergy loss in compressor (kW)
$\dot{E}_{fan}$	Rate of exergy loss in fan (kW)
EV	Expansion valve
h	Enthalpy of humid air (kJ/kg dry air)
$\dot{m}$	Mass flow rate of air stream (kg/s)
$\dot{m}_p$	Mass flow rate of the process air flow (kg/s)
$\dot{m}_r$	Mass flow rate of the reactivation air flow (kg/s)
Q_{cc}	Cooling capacity of cooling coil (kW)
$\dot{Q}_{evop}$	Evaporator energy transfer rate (kW)
$\dot{Q}_{reg}$	Reactivation energy supply (kW)
s	Entropy of humid air (kJ/K kg dry air)
$\dot{S}_{gen}$	Entropy generation (kW/K)
T	Dry bulb temperature (°C)
V	Flow rate of air stream (m³/hr)
VCR	Vapour compression refrigeration

Greek letters

η	Exergy efficiency
ε	Effectiveness
ω	Humidity ratio

Subscripts

0	Dead state
1,2, etc.	State numbers
CC	Cooling coil
DW	Desiccant wheel
HRW	Heat recovery wheel
HT	Heater
p	Process air
r	Regeneration air

Chapter 11

Analysis of Optimum Operating Parameters for Ground Source Heat Pump System for Different Cases of Building Heating and Cooling Mode Operations

T. Sivasakthivel, Vikas Verma, Rahul Tarodiya, Chandan Swaroop Meena, Varun Pratap Singh, and Rajesh Kumar

CONTENTS

11.1 Introduction 183
11.2 Methodology 185
    11.2.1 Taguchi Method 187
    11.2.2 Taguchi – Parameters and Level 187
    11.2.3 GHX Length and Parameters Calculation 192
11.3 Results and Discussions 194
    11.3.1 Effect of Parameters on GHX Length 194
    11.3.2 S/N Ratio.. 198
        11.3.2.1 Space Cooling and Heating 199
        11.3.2.2 Heating < Cooling, Heating = Cooling and Heating > Cooling 202
        11.3.2.3 Thermal Borehole Resistance..................... 206
    11.3.3 ANOVA and Main Effect Analysis 207
        11.3.3.1 Space Heating and Cooling 208
        11.3.3.2 Heating<Cooling, Heating=Cooling, Heating >Cooling 208
        11.3.3.3 Thermal Borehole Resistance..................... 210
    11.3.4 Selection of Optimal Levels and Confirmation Test..... 213
11.4 Conclusions... 213
References 214

11.1 INTRODUCTION

Earth energy is one of the green and sustainable energy sources that can be utilized economically with the help of a ground source heat pump (GSHP) system [1–5]. In a GSHP system, thermal design of a ground heat exchanger (GHX) is an important step for the high performance of GSHP because if it

DOI: 10.1201/9781003395768-11

is not able to extract heat from the ground or reject heat into the ground, then the COP of the GSHP system drops drastically. Hence, for the successful operation of a GSHP system, a suitable GHX has to be selected to meet the required cooling and the heating load [6–9]. The performance of a GHX system depends on many other parameters such as ground properties, mass flow rate, distance between U-pipes, GHX type, diameter of borehole, GHX material, working fluid, pipe diameter, etc. [7–10]. Hence, optimization of these parameters is an important step to improve the thermal performance and economic efficiency of GHX.

Optimization of GSHP and GHX is a wide research area and many optimization models are proposed [6–29]. Some are focused on the control and operational strategy of GSHP system [6,7,9,10,12,13,16]. Gao et al. [18] optimized the control strategy of an air-conditioning integrated GSHP system and obtained reductions of energy consumption in the summer. Gang et al. [19] proposed suitable operating strategies to control a hybrid energy system consisted of GSHP and a cooling tower. An artificial neural network (ANN) was used to predict the dynamic variation in the fluid temperature of a ground loop. Fannou et al. [20] carried out the modelling and optimization of DXGSHP by using ANN for enhancing the efficiency of the system. They found that in the heat pump system, condenser and evaporator play a main important role in meeting the optimum performance of the system. Zhou et al. [21] optimized the performance of a GSHP system through TRNSYS simulation. Sivasakthivel et al. [22] proposed the optimum parameters of a GSHP system for both cooling and heating operations. Zeng et al. [23] used the multi-population genetic algorithm (MPGA) to optimize the operation and capacity strategy for a CCHP–GSHP coupling system, resulting in excellent energy-saving and cost-saving effects in a case study. Esen and Turgut [24] used Taguchi method to optimize the GSHP system and concluded that the depth of borehole is the most significant parameter.

Even though many studies have been conducted with the aim of investigating and evaluating GSHP, still more studies are needed to understand the performance of GHX when GSHP is used for both cooling and heating applications. Some researchers [8,10,11,13,14,17] carried out optimization of a horizontal and vertical GSHP system to reduce the cost of the system. Gang and Wang [25] applied static and dynamic models of ANN to predict the outlet temperature of GHX. They concluded that compared to the static model; the dynamic model was able to predict the exit temperature accurately. Bazkiaei et al. [26] presented a methodology to optimize a horizontal GHX assuming homogeneous and non-homogeneous soil profiles to achieve maximum heat extraction and rejection rates. The performance of GHX was found to be better with a non-homogeneous soil profile than the GHX installed in a homogenous soil profile. Adamovsky et al. [27] analyzed the soil temperature, heat flow and energy transferred from the liner and slinky types of horizontal heat exchangers for heating applications. They concluded that the linear heat exchanger was very efficient and could maintain higher average ground temperature compared to the slinky heat exchanger. Optimization of a vertical ground heat exchanger based on the response surface methodology has been done by Khalajzadeh et al. [28]. Fujii et al. [29] developed an approach for estimating the heat exchange performance of a large-scale GSHP system in Akita Plain, northern Japan, based on a field-wide groundwater flow and heat transport model that had been calibrated against measured data. Sivasakthivel et al. [30] applied Taguchi and utility methods into the optimization of GHX parameters.

In terms of optimizing the operating parameters of GHX, Said et al. [31] presented an optimization methodology for horizontal GHXs based on energy conservation and heat transfer principles. Modelling of a slinky loop GHX was carried out by Esen et al. [32] using ANN and adaptive neuro-fuzzy interface system. Verma and Murugesan [33] carried out optimization of the length of GHX and area of collector for better COP of SAGSHP system. They estimated the reduction in the solar collector area and the length of GHX was around 2.5% and 1.6%, respectively. Wang et al. [34] studied the effect of groundwater flow on the heat transfer process in a spiral coil pile GHX. They concluded that heat transfer rate can be increased by an increase in groundwater flow. When conducting an optimization that determines how much heat should be generated by GHX, it needs to compare many combinations of operations. Due to this, optimization problems that involve GHX along with a GSHP operation cannot be solved easily. One of the best methods is Taguchi optimization. This technique is used in many applications like civil engineering, manufacturing, thermal engineering, etc. [35–41]. In this chapter, an attempt will be made to apply Taguchi technique to optimize the GHX parameters for ground source heat pump applications.

A detailed survey shows that a good number of studies have been carried out on the performance and optimization of the GSHP system [6–34]. The role of GHX which transports heat also needs to be studied to design optimum parameters for different modes of operations. To get the most extreme advantage from GSHP, the framework must be outlined as thermally productive and monetarily moderate. All things are essential to consider the information about the impact of different working parameters for given burden conditions. In given working conditions, optimum working parameters of GHXs need to be analyzed to achieve an optimum coefficient of performance. Hence, there should be a methodology to investigate the required optimum length of the GHX for both space cooling and heating with the variation of cooling and heating periods.

11.2 METHODOLOGY

Experimental analysis of different parameter effects on the performance of the GSHP is a complex analysis; Figure 11.1 show the parameters classification that may affect the performance of the GHX [29,30].

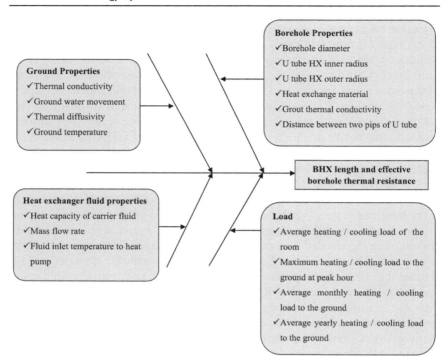

Figure 11.1 Cause and effect diagram for a borehole heat exchanger of GSHP system.

Whenever the performance of the GSHP is improved by considering large number of parameters the number of experiments to be carried out also increases, Therefore, the factors affecting the performance should be determined and checked under laboratory conditions these studies are called as an offline quality improvement. To avoid the large number of experimental trial runs and to consider the more parameters effects on GSHP design, the experimental trial run is an important aspect, design of experiments is one such tool under the principle of design of experiments the number of experimental trial runs need to be carried are reduced and also it presents the interaction between factors. In that way, Taguchi's design of the experiment is a very good tool for analyzing the GSHP system. Taguchi method has been used in manufacturing industries for many years to check the quality of the product [35–41] to reduce the production cost and increase the profit and also in the conventional design of experiment variations caused by uncontrollable factors are not controlled, but in the case of Taguchi method, it allows. From detailed literature, Taguchi method can be applied successfully to thermal engineering problems, so in this chapter, an attempt has been made to apply the Taguchi technique to optimize the ground heat exchanger parameters for ground source heat pump applications.

11.2.1 Taguchi Method

Taguchi optimization is an experimental optimization technique that uses standard orthogonal arrays for forming the matrix of experiments [35–37]. By using this matrix, it will help us to get maximum information from a minimum number of experiments and the best level of each parameter can be found. The major steps of implementing the Taguchi method are presented in Figure 11.2. In data analysis, signal-to-noise (S/N) ratios are used to calculate the response. There are three types of performance characteristics used for analyzing S/N ratios: lower the better, higher the better and nominal the better. The use of ANOVA (analysis of variance) is to find out the percentage contributions of each parameter [35–37,40].

11.2.2 Taguchi – Parameters and Level

In the present analysis, six parameters are considered one is considered at six levels and the other five parameters are considered at three levels. The factors to be studied are mentioned in Table 11.1.

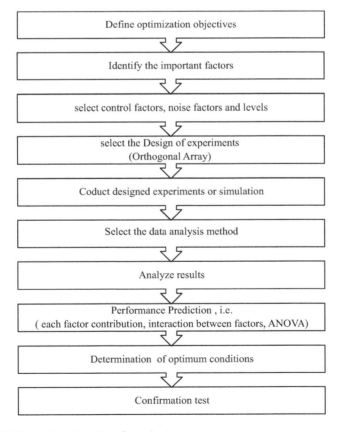

Figure 11.2 Taguchi optimization flow chart.

188 Thermal Energy Systems

Table 11.1 Optimization factors and levels

Label	Factors	Levels					
		1	2	3	4	5	6
A	Mass flow rate of heat carrier fluid per kW (kg/s/kW)	0.04	0.05	0.06	0.07	0.08	0.09
B	Radius of borehole (m)	0.0635	0.0762	0.0889			
C	GHX inner radius (m)	0.0127	0.0159	0.0191			
D	Conductivity of grouting material (W/m/K)	0.73	2.10	2.59			
E	Carrier fluid heat capacity (J/kg/ K)	3991	4187	4288			
F	Distance between two U-tubes (m)	0.05	0.06	0.07			

Optimum Parameters for GSHP System   189

Before selecting an orthogonal array, the least number of experiments to be conducted can be fixed by using the following relation [39].

$$N_{Taguchi} = 1 + NV(L - 1) \tag{11.1}$$

where $N_{Taguchi}$ is the number of experiments to be conducted, NV is the number of variables and L is the number of levels. The standard orthogonal arrays available are L4, L8, L9, L12, L16, L18, etc. According to the Taguchi design concept L18 orthogonal array is chosen for our computations and the structure of the array is shown in Table 11.2.

Each trial run is performed as per the standard L18 array table. The optimization of the observed values is determined by using the signal-to-noise (S/N) ratios and analysis of variance (ANOVA). Tables 11.3, 11.4 and 11.5 show S/N ratio values for heating/cooling only case, heating less than cooling, heating higher than cooling, heating equal to cooling cases and the thermal resistance of the GHX for the L18 array and it is calculated based on the concept of lower the better.

Table 11.2 L_{18} orthogonal array

Experiment no.	Parameters					
	A	B	C	D	E	F
1	1	1	1	1	1	1
2	1	2	2	2	2	2
3	1	3	3	3	3	3
4	2	1	1	2	2	3
5	2	2	2	3	3	1
6	2	3	3	1	1	2
7	3	1	2	1	3	2
8	3	2	3	2	1	3
9	3	3	1	3	2	1
10	4	1	3	3	2	2
11	4	2	1	1	3	3
12	4	3	2	2	1	1
13	5	1	2	3	1	3
14	5	2	3	1	2	1
15	5	3	1	2	3	2
16	6	1	3	2	3	1
17	6	2	1	3	1	2
18	6	3	2	1	2	3

Table 11.3 GHX length with S/N ratio for heating and cooling modes

Experiment no.	Heating mode						Cooling mode					
	20% decrease in average load (7 kW)		average load (8.75 kW)		20% increase in average load (10.5 kW)		20% decrease in average load (7 kW)		average load (8.75 kW)		20% increase in average load (10.5 kW)	
	BHX length (m)	Signal-to-noise ratio (dB)	BHX length (m)	Signal-to-noise ratio (dB)	BHX length (m)	Signal-to-noise ratio (dB)	BHX length (m)	Signal-to-noise ratio (dB)	BHX length (m)	Signal-to-noise ratio (dB)	BHX length (m)	Signal-to-noise ratio (dB)
1	156.1	−43.86	195.2	−45.80	234.2	−47.39	89.6	−39.04	112	−40.98	134.4	−42.56
2	90.7	−39.15	113.4	−41.09	136.1	−42.67	53.5	−34.56	66.9	−36.50	80.3	−38.09
3	82.3	−37.87	98.9	−39.81	119.5	−41.40	49.4	−33.33	62.2	−35.26	74.8	−36.85
4	97.9	−39.81	122.4	−41.75	146.9	−43.34	57	−35.11	71.2	−37.04	85.4	−38.62
5	92	−39.27	115	−41.21	138	−42.79	53.2	−34.51	66.5	−36.45	79.8	−38.04
6	155.2	−43.81	193.9	−45.75	232.7	−47.33	86.6	−38.75	108.2	−40.68	129.9	−42.27
7	143.3	−43.12	179.1	−45.06	214.9	−46.64	79.8	−38.04	99.7	−39.97	119.7	−41.56
8	87.4	−38.83	109.3	−40.77	131.2	−42.35	50.5	−34.06	63.1	−36.00	75.8	−37.59
9	103	−40.25	128.7	−42.19	154.5	−43.77	58	−35.26	72.5	−37.20	87	−38.79
10	86.9	−38.78	108.7	−40.72	130.4	−42.30	50.1	−33.99	62.7	−35.94	75.2	−37.52
11	169.7	−44.59	212.1	−46.53	254.5	−48.11	92.1	−39.28	115.2	−41.22	138.2	−42.81
12	105	−40.42	131.3	−42.36	157.5	−43.94	58.7	−35.37	73.4	−37.31	88.1	−38.89
13	91.3	−39.20	114.1	−41.14	137	−42.73	52.2	−34.35	65.2	−36.28	78.3	−37.87
14	164.5	−44.32	205.7	−46.26	246.8	−47.84	88.9	−38.97	111.1	−40.91	133.4	−42.50
15	112.8	−41.04	141	−42.98	169.2	−44.56	62	−35.84	77.5	−37.78	93.1	−39.37
16	98.2	−39.84	122.8	−41.78	147.3	−43.36	55.2	−34.83	69	−36.77	82.8	−38.36
17	103.9	−40.33	129.9	−42.27	155.9	−43.85	57.8	−35.23	72.3	−37.18	86.7	−38.76
18	173.2	−44.77	216.4	−46.70	259.7	−48.28	92.4	−39.31	115.5	−41.25	138.6	−42.83

Table 11.4 GHX length with their S/N ratio for three different cases of heating and cooling modes

Experiment no.	Heating load less than cooling load				Heating load higher than cooling load				Heating load equal to cooling load			
	Heating BHX length (m)	Signal-to-noise ratio (dB)	Cooling BHX length (m)	Signal-to-noise ratio (dB)	Heating BHX length (m)	Signal-to-noise ratio (dB)	Cooling BHX length (m)	Signal-to-noise ratio (dB)	Heating BHX length (m)	Signal-to-noise ratio (dB)	Cooling BHX length (m)	Signal-to-noise ratio (dB)
1	156.1	-43.86	112	-40.98	195.2	-45.80	89.6	-39.04	234.2	-47.39	134.4	-42.56
2	90.7	-39.15	66.9	-36.50	113.4	-41.09	53.5	-34.56	136.1	-42.67	80.3	-38.09
3	82.3	-37.87	62.2	-35.26	98.9	-39.81	49.4	-33.33	119.5	-41.40	74.8	-36.85
4	97.9	-39.81	71.2	-37.04	122.4	-41.75	57	-35.11	146.9	-43.34	85.4	-38.62
5	92	-39.27	66.5	-36.45	115	-41.21	53.2	-34.51	138	-42.79	79.8	-38.04
6	155.2	-43.81	108.2	-40.68	193.9	-45.75	86.6	-38.75	232.7	-47.33	129.9	-42.27
7	143.3	-43.12	99.7	-39.97	179.1	-45.06	79.8	-38.04	214.9	-46.64	119.7	-41.56
8	87.4	-38.83	63.1	-36.00	109.3	-40.77	50.5	-34.06	131.2	-42.35	75.8	-37.59
9	103	-40.25	72.5	-37.20	128.7	-42.19	58	-35.26	154.5	-43.77	87	-38.79
10	86.9	-38.78	62.7	-35.94	108.7	-40.72	50.1	-33.99	130.4	-42.30	75.2	-37.52
11	169.7	-44.59	115.2	-41.22	212.1	-46.53	92.1	-39.28	254.5	-48.11	138.2	-42.81
12	105	-40.42	73.4	-37.31	131.3	-42.36	58.7	-35.37	157.5	-43.94	88.1	-38.89
13	91.3	-39.20	65.2	-36.28	114.1	-41.14	52.2	-34.35	137	-42.73	78.3	-37.87
14	164.5	-44.32	111.1	-40.91	205.7	-46.26	88.9	-38.97	246.8	-47.84	133.4	-42.50
15	112.8	-41.04	77.5	-37.78	141	-42.98	62	-35.84	169.2	-44.56	93.1	-39.37
16	98.2	-39.84	69	-36.77	122.8	-41.78	55.2	-34.83	147.3	-43.36	82.8	-38.36
17	103.9	-40.33	72.3	-37.18	129.9	-42.27	57.8	-35.23	155.9	-43.85	86.7	-38.76
18	173.2	-44.77	115.5	-41.25	216.4	-46.70	92.4	-39.31	259.7	-48.28	138.6	-42.83

192 Thermal Energy Systems

Table 11.5 Thermal resistances of BHX

Trial no.	Thermal resistance (m.K/W)	Signal-to-noise ratio (dB)
1	0.224	12.99504
2	0.100	20
3	0.082	21.72372
4	0.096	20.35458
5	0.094	20.53744
6	0.226	12.91783
7	0.173	15.23908
8	0.082	21.72372
9	0.117	18.63628
10	0.066	23.60912
11	0.226	12.91783
12	0.118	18.56236
13	0.072	22.85335
14	0.212	13.47328
15	0.127	17.92393
16	0.081	21.8303
17	0.102	19.828
18	0.231	12.72776

11.2.3 GHX Length and Parameters Calculation

GHX is the primary part of a GSHP system which uses ground as a source and hence the performance of GHX is important in GSHP operation. Designers of vertical geothermal systems often need to quickly estimate the total length of a bore field for a given building. One way to do this calculation is to use the sizing equation proposed by Kavanaugh and Rafferty [41] and contained in the *ASHRAE Handbook* [42]. This equation has been recast by Bernier [43] into the following form:

$$L = \frac{q_h R_b + q_y R_{10y} + q_m R_{1m} + q_h R_{6h}}{T_m - (T_g + T_p)} \tag{11.2}$$

L is the total length of the vertical GHX, T_m is the mean fluid temperature in the GHX, T_g is the undisturbed ground temperature, T_p is the temperature penalty ($T_p = 0$, for single BHX), q_y, q_m, q_h represents the yearly average ground load, the highest monthly ground load and the peak hourly ground load respectively. R_{10y}, R_{1m} and R_{6h} are the ground thermal resistance for 10 years, 1 month and six hours, respectively. R_b is the effective GHX thermal resistance. A typical vertical GHX is shown in Figure 11.3. The GHX consists of a single U-tube heat exchanger and borehole filled with grout.

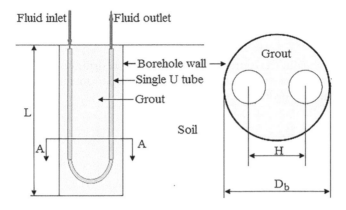

Figure 11.3 Schematic diagram of (a) vertical GHX (b) cross view.

Equation (11.2) was derived assuming that heat transfer in the ground occurs only by conduction and that moisture evaporation or underground water movement are not significant. The effective ground thermal resistances account for transient heat transfer from the borehole wall to the far-field undisturbed ground temperature. Several ways exist to check thermal resistances in the ground. In this work, the approach proposed by the *ASHRAE Handbook* is used [42].

$$R_{6h} = \frac{1}{k} G(\alpha t_{6h}/r_{bore}^2) \tag{11.3}$$

$$R_{1m} = \frac{1}{k} [G(\alpha t_{1m+6h}/r_{bore}^2) - G(\alpha t_{6h}/r_{bore}^2)] \tag{11.4}$$

$$R_{10y} = \frac{1}{k} [G(\alpha t_{10y+1m+6h}/r_{bore}^2) - G(\alpha t_{1m+6h}/r_{bore}^2)] \tag{11.5}$$

where G-function represents the cylindrical heat source solution, k is the ground thermal conductivity, α is the ground thermal diffusivity and r_{bore} is the borehole radius.

For the efficient working of the GHX in two modes, the estimation of the optimum length of GHX is a deciding factor. This is because a longer length will reduce the economic viability of the GHX and a smaller length will reduce the performance of the GSHP system. Taguchi method is one of the efficient optimization techniques to investigate the role of important control factors to achieve an optimum value for a particular objective function.

Thermal resistance R_b is the resistance between the borehole wall and fluid in the pipe and it can be calculated by the following equation [43,44]:

194 Thermal Energy Systems

$$R_b = R_g + \frac{R_p + R_{conv}}{2} \tag{11.6}$$

where R_g is grout resistance, R_p is the pipe resistance and R_{conv} is the convective resistance inside each tube and these parameters can be analyzed by following equations [43,44].

$$R_{conv} = \frac{1}{2\pi r_{p,in} h_{conv}} \tag{11.7}$$

$$R_p = \frac{\ln(r_{p,ext}/r_{p,in})}{2\pi k_{pipe}} \tag{11.8}$$

$$R_g = \frac{1}{4\pi k_{grout}} \left[\ln\left(\frac{r_{bore}}{r_{p,ext}}\right) + \ln\left(\frac{r_{bore}}{H}\right) + \frac{k_{grout} - k}{k_{grout} + k} \ln\left(\frac{r_{bore}^4}{r_{bore}^4 - \left(\frac{H}{2}\right)^4}\right) \right] \tag{11.9}$$

where h_{conv} is the film convection coefficient, $r_{p,in}$ and $r_{p,ext}$, are the inner and outer radius of the U-pipe, k_{pipe} is the thermal conductivity of the pipe material, k_{grout} is the thermal conductivity of the grout, k is the ground thermal conductivity, H is the centre-to-centre distance between the two pipes, r_{bore} is the borehole radius.

11.3 RESULTS AND DISCUSSIONS

Designing a GSHP requires knowledge of ground properties, water movement, types of GHX, radius of borehole and distance between the two U-tube pipes [28,30,34,45]. Among these parameters, some parameters are controllable, some are not. In this chapter, we considered some important parameters which will affect the performance of the GSHP and as well as the heat exchanger length. To carry out this, Taguchi L18 orthogonal array is used. The computational experimental plan of the L_{18} array is presented in Table 11.2. The main aim of this study is to know about different parameter effects on GHX length and thermal resistance of the GHX and to know the effect of the varying load of the room in GHX performance.

11.3.1 Effect of Parameters on GHX Length

The schematic diagram of the GHX is presented in Figure 11.3. To study the effect of different parameters on the GHX length important controllable parameters are considered that are, mass flow rate of the heat carrier fluid flowing in the GHX, radius of the borehole, inner radius of the U-tube heat exchanger, heat capacity of carrier fluid, grout thermal conductivity, centre-to-centre distance between two U-pipes and the values of other controllable and uncontrollable parameters are taken as constant that are thermal conductivity of soil as 2.25 W/m.K, thermal diffusivity of soil as 0.086 m^2/day, ground temperature as 17°C, entering water temperature as 40°C and U-tube heat exchanger thermal conductivity as 0.42 W/m.K. Based on this, the effect of six important parameters is calculated and presented in Figures 11.4–11.9.

Figures 11.4 and 11.5 present the effect of borehole radius and U-tube heat exchanger inner radius on the length of the heat exchanger for different loads. When the borehole radius increases the length required for a given load is increased, but the increase in length is very small in the case of for 7 kW load the length is increased by only 2.2 m. If the U-tube heat exchanger radius increases the length of the heat exchanger is decreased, for 14 kW load, when the radius is increased from 0.0127 to 0.0191 m the length required is decreased by 16 m.

Figures 11.6 and 11.7 illustrate the effect of a centre-to-centre distance between pipes of U-tube and heat capacity of the carrier fluid on the length of the heat exchanger. From Figure 11.6, it is clear that when the distance increases the heat exchanger length is decreasing, but this effect is more when the load is high. In GHX, due to the high-temperature difference between the up and down fluid in adjacent pipes, heat may be transferred directly from one pipe to the other along with the ground. This inter-tube

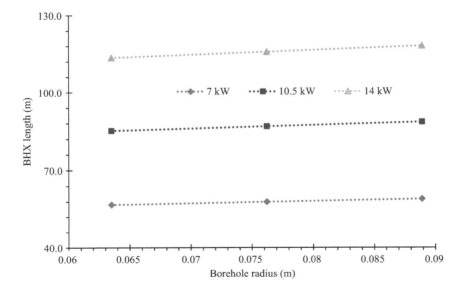

Figure 11.4 Borehole radius effects on BHX length.

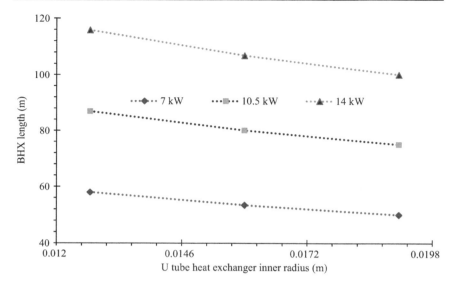

Figure 11.5 U-tube heat exchanger inner radius effects on BHX length.

heat transfer is called thermal 'short-circuit'. It is proportional to the temperature difference between the upward and downward-flowing fluids at a given depth. It is highest at the top of the borehole where the fluid temperature difference is high and low at the bottom. The thermal short-circuit phenomena can reduce heat transfer between the heat carrier fluid and ground, and deteriorate the performance of the GHXs, so the centre distance between pipes of U-tube needs to be maintained carefully.

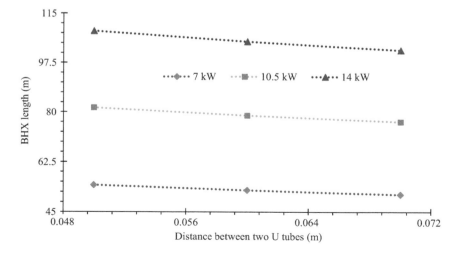

Figure 11.6 Centre-to-centre distance between U-tubes effects on BHX length.

Optimum Parameters for GSHP System 197

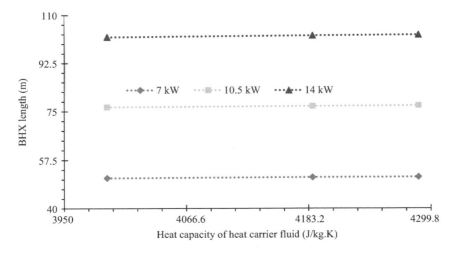

Figure 11.7 Heat capacity of carrier fluid effects on BHX length.

The heat capacity of the carrier fluid does not much affect the length of the heat exchanger, for the case of 10.5 kW when the heat capacity is increased from 3991 to 4288 J/kg.K, the length of the heat exchanger increases from 76.5 to 77.1 m. Figures 11.8 and 11.9 present the effect of grout thermal conductivity and mass flow rate of the carrier fluid on the length of the heat exchanger. The effect of grout thermal conductivity is far more significant than any other parameters on the length of the heat exchanger, and it is also clear from Figure 11.8 for 7 kW loads when the grout thermal conductivity increased from 0.73 to 2.59 W/m.K, the length of the

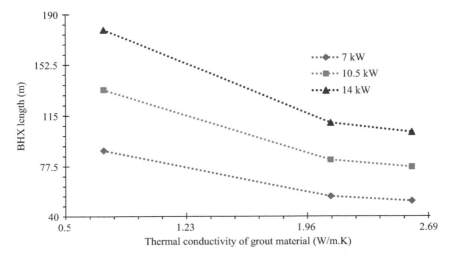

Figure 11.8 Grout thermal conductivity effects on BHX length.

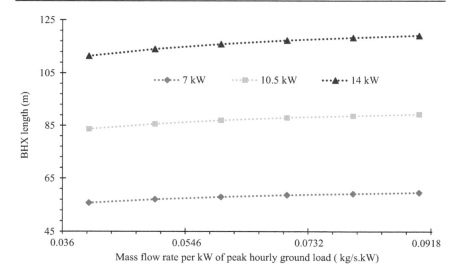

Figure 11.9 Mass flow rate effects on BHX length.

heat exchanger is reduced by 38 m and this effect more significant when the load is increased. When the mass flow rate of carrier fluid increases the length required by the heat exchanger also increases and the effect of mass flow rate for different loads are also presented in Figure 11.9. These different parameters' effect on heat exchanger length is based on keeping some parameters values constant, when all the parameters are varying together then predicting the length of heat exchange becomes a complex analysis, but the effect will be more significant and to undertake such analysis design of experiments become important, based on Taguchi method, these complex analyses are modelled and results of the analysis are presented next.

11.3.2 S/N Ratio

In the Taguchi method once required trial runs are completed next step is to convert the trial results into the signal-to-noise (S/N) ratio. The term signal illustrates the preferable effect for the output (i.e., heat exchanger length and thermal resistance) and the term noise represents the undesirable effects on the heat exchanger length and thermal resistance. S/N ratio gives you the performance characteristics from the desired values. To analyze the Taguchi models, three types of S/N ratios are available that are, smaller the better, larger the better and nominal the best. In this study, the length of GHX and effective thermal resistance analyzed by smaller the better, it is because for given COP and entering water temperature, the length of borehole heat exchange and thermal resistance should be as minimum as possible so that the initial investment of the GSHP will be reduced. Equations (11.10) to (11.12) give the idea about how to calculate S/N ratio for three different cases of Taguchi methods.

For the smaller the better

$$\text{Smaller the best} = -10 * \log_{10}\left(\frac{1}{n}\sum_{i=1}^{n} y_i^2\right) \qquad (11.10)$$

where Y_i is the response, n is the number id tests in a trial; i is the number of repetitions in a trial.

For the larger the better

$$\text{Larger the best} = -10 * \log_{10}\left(\frac{1}{n}\sum_{i=1}^{n} \frac{1}{y_i^2}\right) \qquad (11.11)$$

For the nominal the best

$$\text{Nominal the best} = -10 * \log_{10}\left(\frac{\frac{1}{n}(S_m - V_e)}{V_e}\right)$$

$$\text{where } S_m = \frac{1}{n}\left(\sum_{i=1}^{n} y_i\right)^2 \text{ and } V_e = \frac{1}{n-1}\sum_{i=1}^{n}(y_i - \bar{y})^2 \qquad (11.12)$$

In general, the performance of the system where the smaller the better or larger the better concept are employed are analyzed by the S/N ratio of the system. In our study, we employed smaller the better concept to get the minimum heat exchanger length, in this case, smaller the S/N ratio, smaller the heat exchanger length and less thermal resistance. In nominal the best two-stage optimization process involves one is to maximize the S/N ratio and another is to adjust the output to meet the required optimum.

11.3.2.1 Space Cooling and Heating

The S/N ratio of the only heating and only cooling of a GSHP is presented in Table 11.3 for Taguchi L18 array and for three cases of only heating and only cooling mode. Among these considered cases when the load of the room increases the length of the heat exchanger also increases. The length of the heat exchanger for three different cases of only heating and only cooling is presented in Table 11.3, in the case of only heating, minimum length of the heat exchanger from 18 trial runs are, 98.9 m for average load, 82.3 m for 20% decrease in average load and 119.5 m for a 20% increase in average load. Only cooling case minimum the length of the heat exchanger from 18 trial runs are 62.2 m for average load, 49.4 m for 20% decrease in average load and 74.8 m for a 20% increase in average load. The average values of S/N ratios for the considered six parameters are presented in Table 11.6. The

Table 11.6 Response table for heating and cooling modes of operation

	Heating load	Levels						Delta (S/N_{max} - S/N_{min})	Rank	Overall optimum
		1	2	3	4	5	6			
A	20% decrease	−40.30	−40.97	−40.74	−41.27	−41.53	−41.65	1.35	2	A1
	base load	−42.24	−42.91	−42.68	−43.21	−43.46	−43.59			
	20% increase	−43.82	−44.49	−44.26	−44.79	−45.05	−45.17			
B	20% decrease	−40.77	−41.08	−41.37				0.59	4	B1
	base load	−42.71	−43.02	−43.30						
	20% increase	−44.30	−44.61	−44.89						
C	20% decrease	−41.65	−40.99	−40.58				1.07	3	C3
	base load	−43.59	−42.93	−42.52						
	20% increase	−45.17	−44.51	−44.10						
D	20% decrease	−44.08	−39.85	−39.29				4.79	1	D3
	base load	−46.02	−41.79	−41.23						
	20% increase	−47.60	−43.38	−42.81						
E	20% decrease	−41.08	−41.18	−40.96				0.22	6	E3
	base load	−43.02	−43.12	−42.90						
	20% increase	−44.60	−44.71	−44.48						
F	20% decrease	−41.33	−41.04	−40.85				0.48	5	F3
	base load	−43.27	−42.98	−42.79						
	20% increase	−44.85	−44.56	−44.37						

Heating only

	Cooling load	levels						Delta (S/N_{max} - S/N_{min})	Rank	Overall optimum
		1	2	3	4	5	6			
A	20% decrease	−35.65	−36.13	−35.79	−36.22	−36.39	−36.46	0.82	3	A1
	base load	−37.59	−38.06	−37.73	−38.16	−38.33	−38.40			
	20% increase	−39.17	−39.65	−39.32	−39.74	−39.92	−39.99			
B	20% decrease	−35.90	−36.11	−36.31				0.42	5	B1
	base load	−37.84	−38.05	−38.25						
	20% increase	−39.42	−39.63	−39.84						
C	20% decrease	−36.63	−36.03	−35.66				0.97	2	C3
	base load	−38.57	−37.96	−37.60						
	20% increase	−40.16	−39.55	−39.18						
D	20% decrease	−38.90	−34.97	−34.45				4.45	1	D3
	base load	−40.84	−36.91	−36.39						
	20% increase	−42.43	−38.49	−37.97						
E	20% decrease	−36.14	−36.21	−35.98				0.23	6	E3
	base load	−38.08	−38.15	−37.92						
	20% increase	−39.66	−39.73	−39.50						
F	20% decrease	−36.34	−36.07	−35.91				0.43	4	F3
	base load	−38.28	−38.01	−37.85						
	20% increase	−39.86	−39.60	−39.43						

Cooling only

202 Thermal Energy Systems

average S/N ratio is the index of the output; in this study average S/N ratio is calculated using the smaller the better concept.

The effect of each parameter on the average S/N ratio is presented in Table 11.6 and the table gives a difference of the S/N ratio for different cases of heating-only and cooling-only modes. Based on this difference one can rank the parameters. In only heating case if the parameters are ordered based on the ranking then the order will be grout thermal conductivity, mass flow rate per kW of peak hourly ground load, U-tube heat exchanger inner radius, borehole radius, centre-to-centre distance between U-tubes and heat capacity of heat carrier fluid and for cooling only case the order of parameters are grout thermal conductivity, U-tube heat exchanger inner radius, borehole radius, mass flow rate per kW of peak hourly ground load, centre-to-centre distance between U-tubes, borehole radius and heat capacity of the heat carrier fluid. Figure 11.10 shows the S/N ratio graphs for different cases of only heating mode. Figure 11.10 and Table 11.6 help to identify the best levels of the considered parameters for different load conditions.

It is clear from Figure 11.10, all six parameters have effects on the length of the heat exchanger. Most influencing parameters are grout thermal conductivity, mass flow rate, U-tube inner radius and centre-to-centre distance between the pipes. Based on the S/N ratio table and graph the optimum parameters for heating-only mode and cooling-only mode GSHP system is A1B1C3D3E3F3, i.e., mass flow rate per kW of peak hourly ground load is at 0.04 kg/s/kW, borehole radius is at 0.0635 m, U-tube heat exchanger radius is at 0.0191 m, grout thermal conductivity is at 2.59 W/m/K, the heat capacity of heat carrier fluid is at 4288 J/kg/K and the centre-to-centre distance between U-tubes is at 0.07 m. It also concludes that even if the load of the room is increased the optimum level of the parameters is not affected and moreover optimum parameters are not affected by the type of mode (i.e., heating mode or cooling mode).

11.3.2.2 Heating < Cooling, Heating = Cooling and Heating > Cooling

The S/N ratio of the heating less than cooling, heating equal to cooling and heating greater than cooling types of GSHP is presented in Table 11.4 for the L18 array. Considered cases are possible cases in a place where both heating and cooling is required. Among these considered case length of the heat exchanger varies with the load of the building, length of the heat exchanger for three considered three cases are presented in Table 11.4, in case of heating less than cooling, minimum the length of the heat exchanger to meet both heating and cooling of the building is 82.3 m. Heating equal to cooling case, minimum length of the heat exchanger to meet both heating and cooling of the building is 119.5 m and heating greater than cooling case, minimum length of the heat exchanger to meet both heating and cooling of the building is 98.9 m.

Optimum Parameters for GSHP System 203

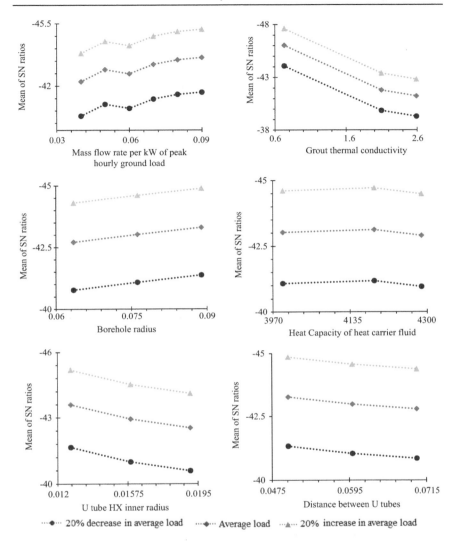

Figure 11.10 Analysis of variance for heating case.

The average values of S/N ratios for the considered six parameters are presented in Table 11.7. The same table helps to identify the optimum levels of considered parameters. It is clear that all six parameters have effects on the length of the heat exchanger. Most influencing parameters are grout thermal conductivity, mass flow rate, U-tube inner radius and centre-to-centre distance between the pipes. Based on the S/N ratio table and graph the optimum parameters for three modes of GSHP system is A1B1C3D3E3F3, i.e., mass flow rate per kW of peak hourly ground load is at 0.04 kg/s/kW, borehole radius is at 0.0635 m, U-tube heat exchanger

Table 11.7 Response table for three different cases of heating and cooling modes of operation

Heating load < cooling load

	Level	A Heating	A Cooling	B Heating	B Cooling	C Heating	C Cooling	D Heating	D Cooling	E Heating	E Cooling	F Heating	F Cooling
Average S/N ratio	1	−40.30	−37.59	−40.77	−37.84	−41.65	−38.57	−44.08	−40.84	−41.08	−38.08	−41.33	−38.28
	2	−40.97	−38.06	−41.08	−38.05	−40.99	−37.96	−39.85	−36.91	−41.18	−38.15	−41.04	−38.01
	3	−40.74	−37.73	−41.37	−38.25	−40.58	−37.60	−39.29	−36.39	−40.96	−37.92	−40.85	−37.85
	4	−41.27	−38.16										
	5	−41.53	−38.33										
	6	−41.65	−38.40										
	Optimum	A1		B1		C3		D3		E3		F3	
	Delta(S/N$_{max}$ − S/N$_{min}$)	1.35	0.82	0.59	0.42	1.07	0.97	4.79	4.45	0.22	0.23	0.48	0.43
	Rank	2	3	4	5	3	2	1	1	6	6	5	4

Heating load > cooling load

	Level	A Heating	A Cooling	B Heating	B Cooling	C Heating	C Cooling	D Heating	D Cooling	E Heating	E Cooling	F Heating	F Cooling
Average S/N ratio	1	−42.24	−35.65	−42.71	−35.90	−43.59	−36.63	−46.02	−38.90	−43.02	−36.14	−43.27	−36.34
	2	−42.91	−36.13	−43.02	−36.11	−42.93	−36.03	−41.79	−34.97	−43.12	−36.21	−42.98	−36.07
	3	−42.68	−35.79	−43.30	−36.31	−42.52	−35.66	−41.23	−34.45	−42.90	−35.98	−42.79	−35.91
	4	−43.21	−36.22										
	5	−43.46	−36.39										
	6	−43.59	−36.46										
	Optimum	A1		B1		C3		D3		E3		F3	
	Delta(S/N$_{max}$ − S/N$_{min}$)	1.35	0.82	0.59	0.42	1.07	0.97	4.79	4.45	0.22	0.23	0.48	0.43
	Rank	2	3	4	5	3	2	1	1	6	6	5	4

Heating load = cooling load

Average S/N ratio

Level	A		B		C		D		E		F	
	Heating	Cooling	Heating	Cooling	Heating	Cooling	Heating	Cooling	Heating	Cooling	Heating	Cooling
1	−43.82	−39.17	−44.30	−39.42	−45.17	−40.16	−47.60	−42.43	−44.60	−39.66	−44.85	−39.86
2	−44.49	−39.65	−44.61	−39.63	−44.51	−39.55	−43.38	−38.49	−44.71	−39.73	−44.56	−39.60
3	−44.26	−39.32	−44.89	−39.84	−44.10	−39.18	−42.81	−37.97	−44.48	−39.50	−44.37	−39.43
4	−44.79	−39.74										
5	−45.05	−39.92										
6	−45.17	−39.99										
Optimum	A1		B1		C3		D3		E3		F3	
Delta(S/N$_{max}$ − S/N$_{min}$)	1.35	0.81	0.59	0.42	1.07	0.97	4.79	4.45	0.22	0.23	0.48	0.43
Rank	2	3	4	5	3	2	1	1	6	6	5	4

206 Thermal Energy Systems

radius is at 0.0191 m, grout thermal conductivity is at 2.59 W/m/K, the heat capacity of heat carrier fluid is at 4288 J/kg/K and the centre-to-centre distance between U-tubes is at 0.07 m.

11.3.2.3 Thermal Borehole Resistance

The S/N ratio of the thermal resistance of the borehole heat exchanger is presented in Table 11.5 for Taguchi L18 array.

The minimum thermal resistance of the borehole heat exchanger from 18 trial runs is 0.066 m.K/W. The average value of S/N ratios of the borehole thermal resistance is presented in Table 11.8. The average S/N ratio is the index of the output; in this study average S/N ratio is calculated using the smaller the better concept. The effect of each parameter on an average S/N ratio is presented in Table 11.8 and also it gives you the difference of the S/N ratio (S/N_{max} - S/N_{min}) for the GHX thermal resistance, based on this difference one can rank parameters, if the parameters are ordered based on the ranking then the order will be grout thermal conductivity, borehole radius, U-tube heat exchanger inner radius, centre-to-centre distance between U-tubes, mass flow rate per kW of peak hourly ground load and heat capacity of the heat-carrying fluid. Figure 11.11 shows the average S/N ratio graphs for different parameters and also it helps to identify the optimum levels of the considered parameters.

It is clear from Figure 11.11 all six parameters have effects on the thermal resistance of borehole heat exchanger. Most influencing parameters are grout thermal conductivity, borehole radius, U-tube heat exchanger inner radius, the centre-to-centre distance between U-tubes. Based on the average S/N ratio response table and graph the optimum parameters for less borehole thermal resistance is A3B1C3D3E3F3, i.e., mass flow rate per kW of peak hourly ground load is at 0.06 kg/s/kW, borehole radius is at 0.0635 m, U-tube heat exchanger radius is at 0.0191 m, grout thermal conductivity is

Table 11.8 Response table for BHX thermal resistance

Level	A	B	C	D	E	F
1	18.24	17.08	19.21	21.20	18.15	18.72
2	18.53	18.08	18.32	20.07	18.36	18.25
3	17.94	19.48	17.11	13.38	18.13	17.67
4	18.36					
5	18.08					
6	18.13					
Delta (S/N_{max} - S/N_{min})	0.60	2.40	2.10	7.82	0.23	1.04
Rank	5	2	3	1	6	4
Optimum	A3	B1	C3	D3	E3	F3

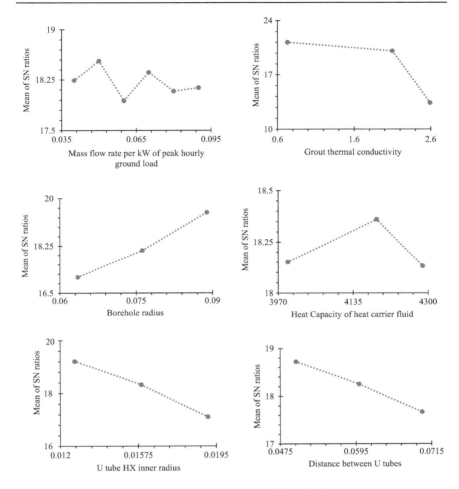

Figure 11.11 Analysis of variance for effective thermal borehole resistance.

at 2.59 W/m/K, the heat capacity of the heat carrier fluid is at 4288 J/kg/K and the centre-to-centre distance between U-tubes is at 0.07 m. It also concludes that in any of the cases, i.e., only heating or only cooling or heating and cooling the optimum level of the parameters are not affected by the load or by the mode of operation.

11.3.3 ANOVA and Main Effect Analysis

ANOVA uses statistical techniques to determine the important parameters that are affecting the length of the borehole heat exchanger and borehole thermal resistance and also it helps to estimate the relative significance of each parameter in terms of percentage contribution to the overall response.

208 Thermal Energy Systems

In this study, analyses were carried out for the confidence level of 95%. ANOVA table contains degrees of freedom, the sum of squares (SS), mean of squares (MS), F ratio, P value and percentage contribution. Among this F ratio is used to find which parameters have a significant effect on the GHX length and resistance. The following formula has been used to calculate the sum of squares (SS), variance and degree of freedom:

$$\text{Correction factor}(C.F) = \frac{\left(\text{sum of } \frac{S}{N}\right)^2}{N} \qquad (11.13)$$

where N is the total number of experiments (N = 18)

$$\text{Degree of Freedom} = \text{Level} - 1 \qquad (11.14)$$

$$\text{Variance} = \frac{SS}{DOF} \qquad (11.15)$$

11.3.3.1 Space Heating and Cooling

ANOVA analysis of three different cases of only heating and only cooling of a GSHP is presented in Tables 11.9 and 11.10 for Taguchi L18 array.

From Table 11.9, it is clear that whenever the load of the room increases from its peak load, the effect of each parameter are at the same level. In the case of heating, we considered three cases of heating load in all of these cases, levels of contribution of each parameter are not affected by the room load pattern. In heating only case important parameters contributions are; grout thermal conductivity (D): 89.77%, mass flow rate per kW of peak hourly ground load (A): 4.24%, U-tube heat exchanger inner radius (C): 3.84, borehole radius (B): 1.16 and factors heat capacity of the heat-carrying fluid (E) and centre-to-centre distance between the U-tubes (F) contribution is around 1% and this can be cross verified by analyzing the P value in Table 11.9. It clearly reveals that the grout thermal conductivity (D) and mass flow rate per kW of peak hourly ground load (A) are the most significant factors (P value < 0.05) affecting both S/N ratio and the length of the GHX.

11.3.3.2 Heating<Cooling, Heating=Cooling, Heating >Cooling

ANOVA analysis for heating less than cooling, heating equal to cooling and heating greater than cooling are presented in Table 11.11.

In all three cases the important parameters that affect the overall performance of GHX are parameter D next to this parameter that affects the performance are parameter A, C, B and parameter E and F are insignificant.

Table 11.9 Analysis of variance for heating mode operation

	20% Decrease in base load					
Parameters	Degree of freedom (DOF)	Sum of squares (SS)	Mean of squares (MS)	F ratio	P value	% Contribution
A	5	3.8914	0.7783	18.66	0.052	4.24
B	2	1.0512	0.5256	12.60	0.074	1.16
C	2	3.5196	1.7598	42.18	0.023	3.84
D	2	82.4250	41.2125	987.91	0.001	89.77
E	2	0.1502	0.0751	1.80	0.357	0.22
F	2	0.7076	0.3538	8.48	0.105	0.77
Error	2	0.0834	0.0417			
Total	17	91.8284				100

	Base load					
Source	DOF	SS	MS	F ratio	P value	% Contribution
A	5	3.8865	0.7773	19.17	0.050	4.24
B	2	1.0408	0.5204	12.84	0.072	1.14
C	2	3.5084	1.7542	43.27	0.023	3.84
D	2	82.3477	41.1739	1015.57	0.001	89.77
E	2	0.1507	0.0753	1.86	0.350	0.23
F	2	0.7120	0.3560	8.78	0.102	0.78
Error	2	0.0811	0.0405			
Total	17	91.7272				100

	20% Increase in base load					
Source	DOF	SS	MS	F ratio	P value	% Contribution
A	5	3.8794	0.7759	18.87	0.051	4.23
B	2	1.0439	0.5220	12.69	0.073	1.13
C	2	3.5146	1.7573	42.74	0.023	3.84
D	2	82.3042	41.1521	1000.79	0.001	89.77
E	2	0.1521	0.0760	1.85	0.351	0.27
F	2	0.7041	0.3520	8.56	0.105	0.76
Error	2	0.0822	0.0411			
Total	17	91.6805				100

The percentage of contribution from parameter D in all the three cases varies from 89% to 92% and the insignificant parameters (E and F) percentage contribution is not more than 1% in all three cases so whenever the GHX is installed for any kind of climatic condition grouting thermal conductivity plays a major role and always one should use the grout which is having high thermal conductivity.

210   Thermal Energy Systems

11.3.3.3 Thermal Borehole Resistance

ANOVA analysis for borehole resistance is presented in Table 11.12. The important parameter that affects the borehole thermal resistance and their percentage contributions are Grout thermal conductivity (D): 85.97%, Borehole radius (B): 6.98%, U-tube heat exchanger inner radius (C): 5.36%, centre-to-centre distance between U-tubes (F): 1.34%, mass flow rate per kW of peak hourly ground load (A): 0.27% and heat capacity of heat carrier fluid (E): 0.08%. The main reason for grouting thermal

Table 11.10 Analysis of variance for cooling mode operation

			20% Decrease in base load			
Parameters	DOF	SS	MS	F ratio	P value	% Contribution
A	5	1.5966	0.3193	7.40	0.123	2.07
B	2	0.5170	0.2585	5.99	0.143	0.63
C	2	2.9028	1.4514	33.63	0.029	3.78
D	2	71.1134	35.5567	823.92	0.001	92.43
E	2	0.1674	0.0837	1.94	0.340	0.43
F	2	0.5550	0.2775	6.43	0.135	0.66
Error	2	0.0863	0.0432			
Total	17	76.9385				100

			Base load			
Source	DOF	SS	MS	F ratio	P value	% Contribution
A	5	1.6019	0.3204	7.36	0.124	2.02
B	2	0.5194	0.2597	5.97	0.144	0.55
C	2	2.9086	1.4543	33.41	0.029	3.78
D	2	71.0627	35.5313	816.26	0.001	92.4
E	2	0.1679	0.0839	1.93	0.342	0.37
F	2	0.5589	0.2794	6.42	0.135	0.88
Error	2	0.0871	0.0435			
Total	17	76.9064				100

			20% Increase in base load			
Source	DOF	SS	MS	F ratio	P value	% Contribution
A	5	1.5980	0.3196	7.19	0.127	2.07
B	2	0.5248	0.2624	5.90	0.145	0.71
C	2	2.8904	1.4452	32.50	0.030	3.75
D	2	71.0953	35.5477	799.40	0.001	92.42
E	2	0.1656	0.0828	1.86	0.349	0.32
F	2	0.5579	0.2790	6.27	0.137	0.73
Error	2	0.0889	0.0445			
Total	17	76.9211				100

Table 11.11 Analysis of variance for three different cases of heating and cooling modes of operation

		Heating load < cooling load									
Parameters	*DOF*	*Sum of squares*		*Mean of squares*		*F ratio*		*P value*		*% Contribution*	
		Heating	*Cooling*	*Heating*	*Cooling*	*Heating*	*Cooling*	*Heating*	*Cooling*	*Heating*	*Cooling*
A	5	3.8914	1.6019	0.7783	0.3204	18.66	7.36	0.052	0.124	4.24	2.02
B	2	1.0512	0.5194	0.5256	0.2597	12.60	5.97	0.074	0.144	1.16	0.55
C	2	3.5196	2.9086	1.7598	1.4543	42.18	33.41	0.023	0.029	3.84	3.78
D	2	82.4250	71.062	1.212	35.531	987.9	816.2	0.001	0.001	89.77	92.4
E	2	0.1502	0.1679	0.0751	0.0839	1.80	1.93	0.357	0.342	0.22	0.37
F	2	0.7076	0.5589	0.3538	0.2794	8.48	6.42	0.105	0.135	0.77	0.88
Error	2	0.0834	0.0871	0.0417	0.0435						
Total	17	91.8284	76.9064							100	100

		Heating load = cooling load									
Parameters	*DOF*	*SS*		*MS*		*F ratio*		*P value*		*% Contribution*	
		Heating	*Cooling*	*Heating*	*Cooling*	*Heating*	*Cooling*	*Heating*	*Cooling*	*Heating*	*Cooling*
A	5	3.8794	1.5980	0.7759	0.3196	18.87	7.19	0.051	0.127	4.23	2.07
B	2	1.0439	0.5248	0.5220	0.2624	12.69	5.90	0.073	0.145	1.13	0.71
C	2	3.5146	2.8904	1.7573	1.4452	42.74	32.50	0.023	0.030	3.84	3.75
D	2	82.3042	71.0953	41.152	35.547	1000.7	799.40	0.001	0.001	89.77	92.42
E	2	0.1521	0.1656	0.0760	0.0828	1.85	1.86	0.351	0.349	0.27	0.32
F	2	0.7041	0.5579	0.3520	0.2790	8.56	6.27	0.105	0.137	0.76	0.73
Error	2	0.0822	0.0889	0.0411	0.0445						
Total	17	91.6805	76.9211								100

(Continued)

212   Thermal Energy Systems

Table 11.11 (Continued) Analysis of variance for three different cases of heating and cooling modes of operation

Parameters	DOF	SS		MS		F ratio		P value		% Contribution	
		Heating	Cooling	Heating	Cooling	Heating	Cooling	Heating	Cooling	Heating	Cooling
A	5	3.8865	1.5966	0.7773	0.3193	19.17	7.40	0.050	0.123	4.24	2.07
B	2	1.0408	0.5170	0.5204	0.2585	12.84	5.99	0.072	0.143	1.14	0.63
C	2	3.5084	2.9028	1.7542	1.4514	43.27	33.63	0.023	0.029	3.84	3.78
D	2	82.3477	71.1134	41.173	35.5567	1015.5	823.92	0.001	0.001	89.77	92.43
E	2	0.1507	0.1674	0.0753	0.0837	1.86	1.94	0.350	0.340	0.23	0.43
F	2	0.7120	0.5550	0.3560	0.2775	8.78	6.43	0.102	0.135	0.78	0.66
Error	2	0.0811	0.0863	0.0405	0.0432						
Total	17	91.7272	76.9385							100	100

Optimum Parameters for GSHP System 213

Table 11.12 Analysis of variance for BHX thermal resistance

Source	DOF	SS	MS	F ratio	P value	% Contribution
A	5	0.678	0.136	19.99	0.048	0.27
B	2	17.417	8.708	1283.98	0.001	6.98
C	2	13.378	6.689	986.23	0.001	5.36
D	2	214.294	107.147	15798.03	0.000	85.97
E	2	0.198	0.099	14.56	0.064	0.08
F	2	3.286	1.643	242.23	0.004	1.34
Error	2	0.014	0.007			
Total	17	249.264				100

conductivity affects more than any other parameters is that it enhances the heat transfer rate between the heat exchanger and the ground.

11.3.4 Selection of Optimal Levels and Confirmation Test

Implementation of the Taguchi method has provided the best optimum parameters. Using these optimum parameters, it is possible to estimate the minimum length and resistance. During analysis, interaction between the parameters is not considered. From the study, it was found that in all the five considered cases the length of GHX after optimization is reduced. In the case of the only heating optimum length of the heat exchanger for 7 kW, 8.75 kW and 10.5 kW heating loads are 77.1 m, 96.3 m and 115.6 m, respectively. Whereas only the cooling case the GHX length for 7 kW 8.75 kW and 10.5 kW loads are 42.7 m, 54.4 m and 66.2 m, respectively. Overall best parameters and levels for all five cases of operation are calculated to be A1B1C3D3E3F3.

11.4 CONCLUSIONS

In this research, a new methodology is introduced to optimize the minimum length required for a GHX by using Taguchi optimization. For this purpose, six operating parameters of the GHX were considered. Conclusions derived from the study are as follows:

- The parametric study shows that whenever the radius of borehole, heat capacity of carrier fluid and mass flow rate per kW increases then the length of the GHX also increases. In the case of the inner radius of U-tube, the distance between tubes and conductivity of grouting material increases means the length of the GHX is decreased.

- In only heating requirement case, the minimum length of GHX required from Taguchi L18 are 82.3 m, 98.9 m and 119.5 m, respectively, for the same cooling load the length of GHXs are 49.4 m, 62.2 m and 74.8 m.
- Taguchi optimization shows that in places where only heating or cooling is required the optimum parameters for heating or cooling will not vary with change in load, i.e., optimum parameters remain the same either the load is 7 or 10.5 kW.
- In all five cases of operation, the best optimum parameters and level are estimated to be A1B1C3D3E3F3.
- Important parameters for GSHP operations are grout conductivity, the flow rate of the heat-exchanging fluid, GHX inner radius, the radius of the borehole, the distance between U-tubes and heat capacity of the heat carrier fluid
- For heating less than cooling, heating equal to cooling and heating higher than cooling cases the optimum length of the GHX are 77.1 m, 115.6 m and 96.3 m, respectively. Thermal resistance of GHX after optimization is 0.062 m.K/W and optimum parameters levels are A3B1C3D3E3F3.

REFERENCES

1. Mustafa OA. Ground-source heat pumps systems and applications. *Renew Sust Energ Rev* 2008;12:344–371.
2. Sarbu I, Sebarchievici C. General review of ground-source heat pump systems for heating and cooling of buildings. *Energy Build* 2014;70:441–454.
3. Zhu N, Hu P, Xu L, Jiang Z, Lei F. Recent research and applications of ground source heat pump integrated with thermal energy storage systems: A review. *Appl Therm Eng* 2014;71:142–151.
4. Sivasakthivel T, Murugesan K, Sahoo PK. A study on energy and CO2 saving potential of ground source heat pump system in India, *Renew Sust Energy Rev* 2014;32:278–293.
5. Luo J, Rohn J, Xiang W, Bertermann D, Blum P. A review of ground investigations for ground source heat pump (GSHP) systems. *Energy Build* 2016;117:160–175.
6. Cui W, Zhou S, Liu X. Optimization of design and operation parameters for hybrid ground-source heat pump assisted with cooling tower. *Energy Build* 2015;99:253–262.
7. Liu W, Chen G, Yan B, Zhou Z, Du H, Zuo J. Hourly operation strategy of a CCHP system with GSHP and thermal energy storage (TES) under variable loads: a case study. *Energy Build* 2015;93:143–153.
8. Gultekin A, Aydin M, Sisman A. Thermal performance analysis of multiple borehole heat exchangers. *Energy Convers Manage* 2016;122:544–551.
9. Capozza A, Zarrella A, De Carli M. Long-term analysis of two GSHP systems using validated numerical models and proposals to optimize the operating parameters. *Energy Build* 2015;93:50–64.

10. Esen H, Inalli M, Esen M. Numerical and experimental analysis of a horizontal ground-coupled heat pump system. *Build Environ* 2007;42:1126–1134.
11. Hu P, Yu Z, Zhu N, Lei F, Yuan X. Performance study of a ground heat exchanger based on the multipole theory heat transfer model. *Energy Build* 2013;65:231–241.
12. Naili N, Hazami M, Attar I, Farhat A. In-field performance analysis of ground source cooling system with horizontal ground heat exchanger in Tunisia. *Energy* 2013;61:319–331.
13. Selamat S, Miyara A, Kariya K. Numerical study of horizontal ground heat exchangers for design optimization. *Renew Energy* 2016;95:561–573.
14. Pahud D, Matthey B. Comparison of the thermal performance of double U-pipe borehole heat exchangers measured in situ. *Energy Build* 2001;33:503–507.
15. Sivasakthivel T, Murugesan K, Sahoo PK. Study of technical, economical andenvironmental viability of ground source heat pump system for Himalayancities of India. *Renew Sust Energy Rev* 2015;48:452–462.
16. Yan L, Hu P, Li C, Yao Y, Xing L, Lei F, et al. The performance prediction of ground source heat pump system based on monitoring data and data mining technology. *Energy Build* 2016;127:1085–1095.
17. Wei J, Wang L, Jia L, Zhu K, Diao N. A new analytical model for short-time response of vertical ground heat exchangers using equivalent diameter method. *Energy Build* 2016;119:13–19.
18. Gao J, Huang G, Xu X. An optimization strategy for the control of small capacity heat pump integrated air-conditioning system. *Energy Convers. Manage.* 119 (2016) 1–13.
19. Gang W, Wang J, Wang S. Performance analysis of hybrid ground source heat pump systems based on ANN predictive control. *Appl Energy* 2014;136:1138–1144.
20. Fannou JLC, Rousseau C, Lamarche L, Kajl S. Modeling of a direct expansion geothermal heat pump using artificial neural networks. *Energy Build* 2014;81:381–390.
21. Zhou S, Cui W, Zhao S, Zhu S. Operation analysis and performance prediction for a GSHP system compounded with domestic hot water (DHW) system. *Energy Build* 2016;119:153–163.
22. Sivasakthivel T, Murugesan K, Thomas HR. Optimization of operating parameters of ground source heat pump system for space heating and cooling by Taguchi method and utility concept. *Appl Energy* 2014;116:76–85.
23. Zeng R, Li H, Liu L, et al. A novel method based on multi-population genetic algorithm for CCHP–GSHP coupling system optimization. *Energy Convers Manage* 2015;105:1138–1148.
24. Esen H, Turgut E. Optimization of operating parameters of a ground coupled heat pump system by Taguchi method. *Energy Build* 2015;107:329–334.
25. Gang W, Wang J. Predictive ANN models of ground heat exchanger for the control of hybrid ground source heat pump systems. *Appl Energy* 2013;112:1146–1153.
26. Bazkiaei AR, Dehghan NE, Kolahdouz EM, Webera AS, Dargush GF. A passive design strategy for a horizontal ground source heat pump pipe operation optimization with a non-homogeneous soil profile. *Energy Build* 2013;61:39–50.

27. Adamovsky D, Neuberger P, Adamovsky R. Changes in energy and temperature in the ground heat exchangers – the energy source for heat pumps. *Energy Build* 2015;92:107- 115.
28. Khalajzadeh V, Heidarinejad G, Srebric J. Parameters optimization of a vertical ground heat exchanger based on response surface methodology. *Energy Build* 2011;43:1288–1294.
29. Fujii H, Itoi R, Fujii J, Uchida Y. Optimizing the design of large-scale ground-coupled heat pump systems using groundwater and heat transport modeling. *Geothermics* 2005;34:347–364
30. Sivasakthivel T, Murugesan K, Sahoo PK. Optimization of ground heat exchanger parameters of ground source heat pump system for space heating applications. *Energy* 2014;78:573–586.
31. Said SAM, Habib MA, Mokheimer EMA, Shayea NAl, Sharqawi M. Horizontal ground heat exchanger design for ground-coupled heat pumps. In Ecologic Vehicles Renewable Energies, (Conference at Monaco). 1–8. 2009.
32. Esen H, Esen M, Ozsolak O. Modelling and experimental performance analysis of solar-assisted ground source heat pump system. *J Exp Theor Artif Intell*2015;19(1): 1–17. 10.1080/0952813X.2015.1056242.
33. Verma V, Murugesan K. Optimization of solar assisted ground source heat pump system for space heating application by Taguchi method and utility concept. *Energy Build* 2014;82:296–309.
34. Wang D, Lu L, Zhang W, Cui P. Numerical and analytical analysis of groundwater influence on the pile geothermal heat exchanger with cast-in spiral coils. *Appl Energy* 2015;160:705–714.
35. Singh H, Kumar P. Optimizing multi-machining characteristics through Taguchi's approach and utility concept. *J Manuf Technol Manage* 2006;17:255–274.
36. Kumar J, Khamba JS. Multi-response optimization in ultrasonic machining of titanium using Taguchi's approach and utility concept. *Int J Manuf Res* 2010;2:139–160.
37. Kumar P, Barua B, Gaindhar JL. Quality optimization (multi-characteristic) through Taguchi's technique and utility concept. *Qual Reliab Eng Int* 2000;16:475–485.
38. Sivasakthivel T, Murugesan K Kumar S, Hu P, Kobiga P. Experimental study of thermal performance of a ground source heat pump system installed in a Himalayan city for composite climatic conditions. *Energy Build* 2016;131:193–206.
39. Rahim A, Sharma UK, Murugesan K, Sharma A, Arora P. Optimization of postfire residual compressive Strength of concrete by Taguchi method. *J Struct Fire Eng* 2012;3:169–180.
40. Ganapathy T, Murugesan K, Gakkhar RP. Performance optimization of Jatropha biodiesel engine model using Taguchi approach. *Appl Energy* 2010;86:2476–2486.
41. Kavanaugh SP, Rafferty K. Ground-source heat pumps: design of geothermal systems for commercial and institutional buildings. *ASHRAE*, Chap. 3; 1997.
42. *ASHRAE Handbook—HVAC* Applications. Chap. 32; 2007
43. Bernier, M. Closed-loop ground-coupled heat pump systems. *ASHRAE J* 2006;48:12–19.

44. Philippe M, Bernier M, Marchio D. Vertical geothermal borefields. *ASHRAE J* 2010;52:20–28.
45. Zhang C, Chen P, Liu Y, Sun S, Peng D. An improved evaluation method for thermal performance of borehole heat exchanger. *Renew Energy* 2015;77:142–151.

NOMENCLATURE

GSHP	Ground source heat pump
ASHP	Air source heat pump
ANN	Artificial neural network
TRT	Thermal response test
GHX	Ground heat exchanger
CF	Correction factor
HP	Heat pump
COP	Coefficient of performance
BHX	Borehole heat exchanger
S/N ratio	Signal-to-noise ratio
ANOVA	Analysis of variance
NV	Number of variables
DOF	Degree of freedom
in	Inlet
out	Outlet
$N_{Taguchi}$	Number of Taguchi trial runs
T	Temperature
L	Levels
G	Cylindrical heat source solution
a	Ground thermal diffusivity
t	Time
h_{conv}	Convective heat transfer coefficient
k_{pipe}	Pipe thermal conductivity
L	Total borehole length
T_m	Mean fluid temperature in the borehole
T_g	Undisturbed ground temperature
T_p	Temperature penalty
r_{bore}	Borehole radius
h_{conv}	Convection coefficient
k	Ground thermal conductivity
$r_{p,in}$	Inner radius of the U pipe
H	Centre-to-centre distance
$r_{p,ext}$	Outer radius of the U-pipe
k_{pipe}	Thermal conductivity of the pipe
f_e	Error in degree of freedom
k_{grout}	Thermal conductivity of the grout

218 Thermal Energy Systems

R_{10y} 10 years ground thermal resistances
R_{1m} One month ground thermal resistances
R_{6h} Six hours ground thermal resistances
q_h Peak hourly ground load
q_m Highest monthly ground load
q_y Yearly average ground heat load
R_b Effective borehole thermal resistance
R_g Effective grout thermal resistance
R_p Effective pipe thermal resistance
R_{conv} Effective convective resistance

Chapter 12

Lithium-Ion Battery Thermal Management Systems with Different Mediums and Techniques for Electric-Driven Vehicles

Saurav Sikarwar, Rajesh Kumar, Ashok Yadav, Nitesh Dutt, Vikas Verma, and T. Sivasakthivel

CONTENTS

12.1 Introduction 219
12.2 Battery Thermal Management System...221
    12.2.1 Air-Based Thermal Management System........................221
    12.2.2 PCM-Based Thermal Management System.....................222
    12.2.3 Liquid-Based Thermal Management System...................223
12.3 Conclusion 225
References..228

12.1 INTRODUCTION

Covid-19 pandemic taught a lesson to human beings one more time that how important environmental conditions are globally [1]. From the end of the 20th century, there are two major concerns, first the lack of energy resources and the second one is environmental pollution. The automobile sector plays a major role in environment contamination [2]. The worldwide fossil fuel consumption rate increased by 1.9 million barrels in a single day, which was approximately two-thirds used by the automobile sector for transportation purposes [3] especially air pollution is most important for humans as well as animals and plant (tree) growth. In that context, the conventional technology is not capable to improve the environmental conditions or to reduce air pollution. Hence, keeping in mind the above crisis we need to switch to other green technologies that release lesser emissions [1]. Therefore, all the countries are moving to words, EVs (electric vehicles) and hybrid EVs (HEVs) which have zero tailpipe emissions. To control the increased air pollution, EVs are useful nowadays [1,4]. Worldwide the government promoted the use of EVs, adding the beneficial policies for EVs [5]. Approximately 315,000 EVs were dispatched from different dealers to sell globally in 2014, which showed a 50% increase in data from the previous year (2013) that data was approximately increased 550,000 in 2015. In the same way, that data increased by approximately

DOI: 10.1201/9781003395768-12

774,000 in the year 2016 which is 40% more than the previous year (2015) [6]. Internal combustion engine (ICE) is the backbone of our transportation system, which gives the best efficiency within the narrow limit of temperature from 90°C to 95°C. In the same way, the battery pack (BP) is the backbone of EVs and HEVs which also gives the best efficiency with a limited range of temperature. As per the literature review, the lithium-ion battery (LiB) is the winner among all batteries for the application of EVs [7]. Furthermore, LiB gives better efficiency within the limited temperature range from 15°C to 35°C [1]. In the automobile sector, EVs work as a green technology. It is important to promote the use of LiBs for transportation vehicles [8]. In the manufacturing of all traction batteries in 2015, the production of LiBs was 28%. The anticipation is that the maximum percentage is expected to increase in the future [9,10]. Heat is generated during the fast running and fast charging of the EV. If there is more heat generation than heat rejection from the BP the temperature of the BP increases. If the BP is not maintained within the optimum range, the performance of BP is reduced [11]. As usual, the temperature rose during the charging and discharging period at that time heat could not be dissipated from the BP. When the temperature increases more than 80°C, thermal runaway occurs and get explosions in BP [12]. This situation is harmful for battery life and the safety of passengers [13,14]. Hence, to maintain the temperature within the optimum range, battery thermal management plays an important role.

Generally, the BTMSs use the air medium, liquid medium and phase change material (PCM) in the application of EVs [15]. As shown in Figure 12.1, from 0°C to 10°C the capacity of the battery pack fades which

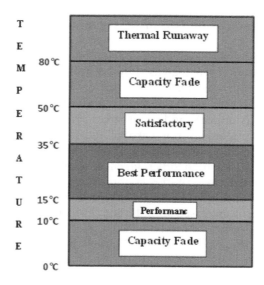

Figure 12.1 Performance of lithium-ion battery with temperature gradient [16].

LiB Thermal Management Systems 221

is due to chemical actions inside the cell. Above 10°C, the performance of LiB is quite better. Further from 15°C to 35°C, it gives the best performance, whereas up to 50°C is the tolerable limit. Furthermore, again it lost capacity and power up to 80°C. If we are not preventing the increases in the temperature rate above 80°C, the BP may catch fire and it is called thermal runaway [16].

12.2 BATTERY THERMAL MANAGEMENT SYSTEM

In conventional technology, ICE cooling systems are mandatory to prevent overheating of the system with regards to the cooling system. Similarly, in the application of EVs or HEVs, it needs to maintain the temperature within a limited range. For this purpose, a battery thermal management system (BTMS) is used in EVs and HEVs [17]. BTMS ensures good battery life and prevents thermal runaway by maintaining the optimum temperature of the battery pack [18]. An electric-driven vehicle BP is constructed with a number of cells, and the cells are interconnected with each other by means of parallel and series connections. Air, liquid and PCM-based BTMS are used as a cooling medium which removes the heat from the system. A lot of research work has been carried out on the air system [17]. All the above systems are used to prevent thermal runaway [18]. Zhang et al. highlighted the following advantages of BTMSs:

1. Heat dissipation with the help of BTMSs at a higher temperature.
2. Heated up when the battery temperature is low to safely charge and discharge the BP.
3. To decrease the cell-to-cell temperature within BP.
4. By default, generated local hot spot in the BP are reduced with the help of BTMS.
5. To prevent, a fast decline in performance at high-temperature conditions [19].

The selection of BTMS is dependent on the rating of BP, if the BP rating is low the air-based BTMS and PCM-based BTMSs are enough to cool BP. If the BP rating is high, the liquid-based and hybrid-type BTMSs are used. The following three types of BTMSs are most popularly used in EVs applications [17].

12.2.1 Air-Based Thermal Management System

In the air-based BTMSs, the air is passed through the BP, which removes the heat from the system, the air serves as a cooling medium in this system. There are two ways of using air as a cooling medium. First is the natural convection cooling method and the second is the forced convection cooling

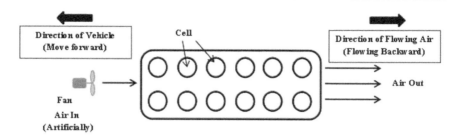

Figure 12.2 Passive (natural convection) air cooling system [1].

method. The cooling fan is used to flow the air forcefully in the later method. Air-based BTMSs require less maintenance. It has a simple design and low cost. Natural convection and forced convection methods in air-based BTMS are also known as passive and active cooling methods, respectively. Figures 12.2 and 12.3 represent passive (natural) air cooling systems and active (artificial) air cooling systems, respectively. The transient output from the BP, i.e., the heat coefficient interrelated by means of Nusselt number and Reynolds number, as is given by equation (12.1) [20].

$$NU = 0.374 \ Re^{0.8014} \qquad (12.1)$$

Although, the air has low thermal conductivity and low heat-removing capability. The air cooling-based BTMSs have limited use for low ampere hours rating BP [22,23].

As shown in Figure 12.2, the vehicle moves in the forward direction while the air moves in the backward direction, which results in the cooling of BP naturally. This is called a passive air cooling, but the passive cooling is suitable for small setups. When a high rate BP is used, there are requirements for cooling fans, which is known as active air cooling [20].

12.2.2 PCM-Based Thermal Management System

There are plenty of works with the use of PCM. The PCM can also be used for the thermal management of BP [24–26]. The PCM cooling

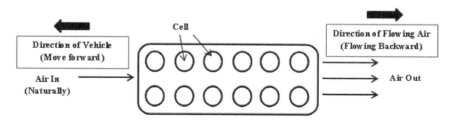

Figure 12.3 Active (forced convection) air cooling system [21].

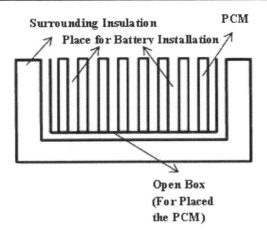

Figure 12.4 PCM-based thermal management system [29].

techniques for BP were investigated by Al-Hallaj and Selman in 2000 which comes under the passive cooling method [27]. PCM cooling technique has many advantages such as compact size, low cost, and no need for pipe networks [20]. Furthermore, the phase changes properties during the operational condition of the battery, it absorbs and releases heat from the BP and maintains the required temperature of BP [28]. As shown in Figure 12.4, PCM is wrapped around the battery cell with a PCM container after that it's covered by an insulation box to exchange heat from the BP [29]. The principle of PCM is shown in Figure 12.5 (graph), PCM works like a heat sink, during the BP discharge and gets a high rate of heat generation and vice-versa. Melting of PCM prevents the temperature rise rapidly in BP. There is no need for any type of external electrical device like fans, pumps as well as manifolds. Although there are a few drawbacks to PCM like during the phase change it requires the space to change the volume. PCM does not regain its original shape if it melts completely. Also, it has a poor thermal conductivity. Therefore, pure PCM has limited use for BP cooling but can be used with some other materials like by mixing a few percentage of graphite can increase the thermal conductivity of PCM [30].

12.2.3 Liquid-Based Thermal Management System

In the liquid-based cooling method, one of the important concerns of water as a cooling medium but in the EVs or HEVs water is dangerous for direct cooling of BP. That is a big issue with direct cooling with water so the EV industries like Tesla and GM used indirect cooling with the medium of water. The more popular coolant mixture of water and glycol in the case of

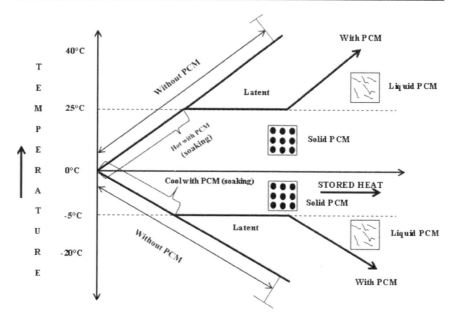

Figure 12.5 Principle of PCM [24,29].

EVs and HEVs for heat transfer (Li Zhu [6]). Liquid cooling is divided into two categories direct/immerse cooling methods and indirect cooling methods, indirect cooling methods can further be classified as tube type, cold plate with fins and jacket-type cooling systems. Another way to classify is on the behalf of cell shape, i.e., cylindrical in shape, pouch type, prismatic type with the different way of heat sinks. The liquid-based BTMSs are shown in Figure 12.6(a) and (b) [31,32].

As shown in Figure 12.6(a), Tesla used the serpentine tubes for cooling the BP. In this design, the cell is partially covered by a tube. Due to the cell's indirect contact with the tube, it is safer by means of electricity in comparison to the coolant jacket design. The serpentine design is also used with fewer welding joints which causes the smooth flow of coolant and greatly reduces the coolant leakage issues. Furthermore, this design is used for less weight and area of BP [6].

Figure 12.6(b) shows the coolant jacket, it has the hollow tunnels for the flow of coolant, the cells arranged through the cavity itself, the cavity also has an inlet and outlet path for a coolant where the coolant enters and exits. The separation of walls guides the coolant flows through contact with the maximum area of the cell indirectly. Even the cavity helps to install the cells. In a comparison of tube-type BTMSs, the coolant jacket is better thermally because the whole surface of the coolant

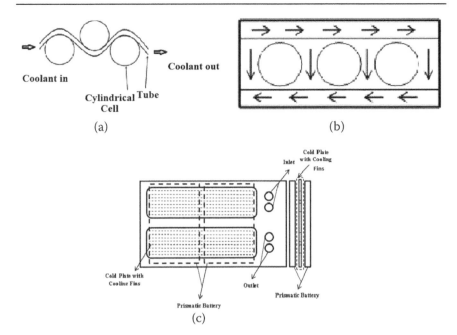

Figure 12.6 (a) Tube-type cooling liquid BTMS [6]. (b) Coolant jacket-type liquid BTMS [31,33]. (c) Cold plate with fins-type Liquid BTMS [6,34–36].

contacts the battery cell with the separation of walls. Through the coolant jacket in different segments, geometry maintained the temperature uniformity [31,33].

A cold plate with fins is also used for battery cooling as shown in Figure 12.6(c). It has many liquid channels with fins to remove the heat from the BP. This design is suitable for prismatic batteries because the cold plate surface is also flat. Hence, it is compatible with prismatic batteries as compared to the other conventional design of BTMSs used with straight channels and a circular section that directly creates resistance to the wall. The oblique construction increases the boundary layer. It is directly affected by the heat removed from the heat sink. Cold plate with fins, type BTMS increases the capacity in comparison to conventional cooling channels with minimal pressure [33–36].

12.3 CONCLUSION

- Approximately 315,000 EVs were dispatched from different dealers to sell globally in 2014 which showed a 50% increase in data from the previous year (2013) that data was approximately increased to

550,000 in 2015. In the same way, that data increased by approximately 774,000 in the year 2016 which is 40% more than the previous year (2015).

- The lithium-ion battery (LiB) is a winner among all traction batteries for the application of EVs. Furthermore, LiB gives higher efficiency within the limited temperature range of 15°C–35°C.
- In the manufacturing of all traction batteries in 2015, the production of LiBs was 28% and the anticipation of that percentage also increased in the future.
- To maintain the temperature within the optimum range, battery thermal management plays an important role. The BTMS generally use the air medium, liquid medium and phase change material (PCM) in the application of EVs currently heat pipe is also used in BTMS for battery cooling.
- From 0°C to 10°C, the capacity fade of the battery pack is due to improper chemical actions inside the cell. Above 10°C, the performance is quite better. From 15°C to 35°C, it gives the best performance, whereas up to 50°C is the tolerable limit. Furthermore, it loses its capacity and power from 50°C to 80°C. If we do not prevent the increase in the temperature, beyond 80°C. Thermal runaway occurs.
- The BTMS is selected on the basis of the power rating of BP for low-rating air-based BTMS and PCM-based BTMS is used. Similarly, for a higher rating, liquid-based and hybrid-type is used.
- In the air-based type BTMSs, the air passed through the BP, to remove the heat from the system, there are two methods of using air first is natural convection cooling (passive cooling) and second is forced convection cooling method (active cooling).
- PCM cooling techniques have plenty of advantages it is compact, low cost, there is no need for pipes such as cooling tubes. As phase changes, the properties of the material also change during the operational condition. It absorbs the heat from the BP and releases heat into the environment due to the phase change process so it maintains the temperature in the operating range.
- Similarly, liquid cooling is divided into two categories direct/immerse cooling method and indirect cooling method, the indirect cooling method is further classified as tube type, cold plate with fins and jacket-type cooling. Another way to classify is on the behalf of cell shape that is cylindrical in the shape, pouch type, prismatic type with the different ways of heat sinks.
- Finally, Table 12.1 shows the comparison of air-based, PCM-based and liquid-based BTMSs. It can be concluded that air and PCM for low-rate traction batteries while the high-rate traction batteries require the liquid-based and hybrid-type BTMSs.

Table 12.1 BTMS summary of medium wise

S.No.	References	Cooling medium	Thermal conductivity	Cooling method	Details	Main components	Attributes	Applicable	Number
I	[6,21,23,37]	Air	Low	Active / passive	Air passed over the BP and removes the heat	Fan, channels	Less maintenance, simple design, low cost, large space required, easy construction, controlled cooling and heating, affected by weather.	Limited but enough cooling for 48 V/12 V EVs and HEVs	The heat coefficient interrelated of NU and Re
2	[25,26,28, 30,38]	PCM	Low	Passive	Absorb and release heat from the BP, maintains the limited temperature	Phase change material	Compact, low cost, there is no need for pipe networks, no need for an electrical device, required the space to change the volume, no weather effects	Testing in laboratory	------
3	[31–36,39]	Liquid	High	Active / passive	Liquid cooling is divided into two categories direct/immerse cooling methods and indirect cooling methods	Cold plate, fins, channels, pipe, pump	Controlled cooling and heating, best performance, no weather effects, applicable for EV, plug-in hybrid electric vehicle (PHEV), HEV	Optimum temperature for 48 V/12 V and higher range of EV, PHEV, HEV	The heat coefficient interrelated of NU and Re

- In the future, to utilize the LiB batteries more efficiently, LiB pack must have more battery backup, fast charging, long trip, no range anxiety, and also very importantly there must be no thermal runaway.

Declarations: The author declares that they have no conflict of interest.

REFERENCES

1. H. Fayaz, *et al.*, *Optimization of Thermal and Structural Design in Lithium-Ion Batteries to Obtain Energy Efficient Battery Thermal Management System (BTMS): A Critical Review*, vol. 29, no. 1. Springer Netherlands, 2022
2. S. F. Tie and C. W. Tan, "A review of energy sources and energy management system in electric vehicles," *Renew. Sustain. Energy Rev.*, vol. 20, pp. 82–102, 2013
3. H. Liu, Z. Wei, W. He, and J. Zhao, "Thermal issues about Li-ion batteries and recent progress in battery thermal management systems: A review," *Energy Convers. Manag.*, vol. 150, pp. 304–330, May 2017
4. A. Afzal, A. D. Mohammed Samee, R. K. Abdul Razak, and M. K. Ramis, "Thermal management of modern electric vehicle battery systems (MEVBS)," *J. Therm. Anal. Calorim.*, vol. 144, no. 4, pp. 1271–1285, 2021
5. M. Van Der Steen, R. M. Van Schelven, R. Kotter, M. Van Twist, and P. Van Deventer Mpa, "EV policy compared: An international comparison of governments' policy strategy towards E-mobility," *Green Energy Technol.*, vol. 203, pp. 27–53, 2015
6. G. Xia, L. Cao, and G. Bi, "A review on battery thermal management in electric vehicle application," *J. Power Sources*, vol. 367, pp. 90–105, August 2017
7. A. Greco, D. Cao, X. Jiang, and H. Yang, "A theoretical and computational study of lithium-ion battery thermal management for electric vehicles using heat pipes," *J. Power Sources*, vol. 257, pp. 344–355, 2014
8. J. Speirs, M. Contestabile, Y. Houari, and R. Gross, "The future of lithium availability for electric vehicle batteries," *Renew. Sustain. Energy Rev.*, vol. 35, pp. 183–193, 2014
9. G. Zhang, L. Cao, S. Ge, C.-Y. Wang, C. E. Shaffer, and C. D. Rahn, "In situ measurement of radial temperature distributions in cylindrical Li-ion cells," *J. Electrochem. Soc.*, vol. 161, no. 10, pp. A1499–A1507, 2014
10. T. M. Bandhauer, S. Garimella, and T. F. Fuller, "A critical review of thermal issues in Lithium-ion batteries," *J. Electrochem. Soc.*, vol. 158, no. 3, p. R1, 2011
11. D. Doughty and E. P. Roth, "A general discussion of Li ion battery safety," *Electrochem. Soc. Interface*, vol. 21, no. 2, pp. 37–44, 2012
12. Z. Ling, *et al.*, "Review on thermal management systems using phase change materials for electronic components, Li-ion batteries and photovoltaic modules," *Renew. Sustain. Energy Rev.*, vol. 31, pp. 427–438, 2014
13. Q. Wang, P. Ping, X. Zhao, G. Chu, J. Sun, and C. Chen, "Thermal runaway caused fire and explosion of lithium ion battery," *J. Power Sources*, vol. 208, pp. 210–224, 2012

14. J. Xu, C. Lan, Y. Qiao, and Y. Ma, "Prevent thermal runaway of lithium-ion batteries with minichannel cooling," *Appl. Therm. Eng.*, vol. 110, pp. 883–890, 2017

15. A. Alrashdan, A. T. Mayyas, and S. Al-Hallaj, "Thermo-mechanical behaviors of the expanded graphite-phase change material matrix used for thermal management of Li-ion battery packs," *J. Mater. Process. Technol.*, vol. 210, no. 1, pp. 174–179, 2010

16. H. Jouhara, *et al.*, "Applications and thermal management of rechargeable batteries for industrial applications," *Energy*, vol. 170, pp. 849–861, 2019

17. L. H. Saw, Y. Ye, and A. A. O. Tay, "Electrochemical–thermal analysis of 18650 Lithium iron phosphate cell," *Energy Convers. Manag.*, vol. 75, pp. 162–174, Nov. 2013

18. J. Liu, H. Li, W. Li, J. Shi, H. Wang, and J. Chen, "Thermal characteristics of power battery pack with liquid-based thermal management," *Appl. Therm. Eng.*, vol. 164, September 2019, p. 114421, 2020

19. D. Chen, J. Jiang, G. H. Kim, C. Yang, and A. Pesaran, "Comparison of different cooling methods for lithium ion battery cells," *Appl. Therm. Eng.*, vol. 94, pp. 846–854, 2016

20. Saurav Sikarwar, Rajesh Kumar, Ashok Yadav, Manoj Gwalwanshi "Battery thermal management system for the cooling of Li-Ion batteries, used in electric vehicles" *Materials Today: Proceedings*, ISSN 2214-7853, 2023

21. M. Al-Zareer, I. Dincer, and M. A. Rosen, "A review of novel thermal management systems for batteries," *International Journal of Energy Research*, vol. 42, no. 10. John Wiley and Sons Ltd, pp. 3182–3205, Aug. 01, 2018

22. T. Wang, K. J. Tseng, J. Zhao, and Z. Wei, "Thermal investigation of lithium-ion battery module with different cell arrangement structures and forced air-cooling strategies," *Appl. Energy*, vol. 134, pp. 229–238, 2014

23. H. Wang, F. He, and L. Ma, "Experimental and modeling study of controller-based thermal management of battery modules under dynamic loads," *Int. J. Heat Mass Transf.*, vol. 103, pp. 154–164, 2016

24. E. M. Shchukina, M. Graham, Z. Zheng, and D. G. Shchukin, "Nanoencapsulation of phase change materials for advanced thermal energy storage systems," *Chem. Soc. Rev.*, vol. 47, no. 11, pp. 4156–4175, 2018

25. A. Sharma, V. V. Tyagi, C. R. Chen, and D. Buddhi, "Review on thermal energy storage with phase change materials and applications," *Renew. Sustain. Energy Rev.*, vol. 13, no. 2, pp. 318–345, 2009

26. L. F. Cabeza, A. Castell, C. Barreneche, A. De Gracia, and A. I. Fernández, "Materials used as PCM in thermal energy storage in buildings: A review," *Renew. Sustain. Energy Rev.*, vol. 15, no. 3, pp. 1675–1695, 2011

27. M. Al-Zareer, I. Dincer, and M. A. Rosen, "Heat and mass transfer modeling and assessment of a new battery cooling system," *Int. J. Heat Mass Transf.*, vol. 126, pp. 765–778, 2018

28. D. Zhao, S. Deng, Y. Shao, L. Zhao, P. Lu, and W. Su, "A new energy analysis model of seawater desalination based on thermodynamics," *Energy Procedia*, vol. 158, pp. 5472–5478, 2019

29. Q. Zhang, A. Agbossou, Z. Feng, and M. Cosnier, "Solar micro-energy harvesting based on thermoelectric and latent heat effects. Part II: Experimental analysis," *Sensors Actuators, A Phys.*, vol. 163, no. 1, pp. 277–283, 2010

30. S. Al-Hallaj, R. Kizilel, A. Lateef, R. Sabbah, M. Farid, and J. Rob Selman, "Passive thermal management using phase change material (PCM) for EV and HEV Li-ion batteries," *IEEE Vehicle Power and Propulsion Conference, Chicago, IL, USA*, p. 5, 2005. 10.1109/VPPC.2005.1554585.

31. J. Zhao, Z. Rao, and Y. Li, "Thermal performance of mini-channel liquid cooled cylinder based battery thermal management for cylindrical lithium-ion power battery," *Energy Convers. Manag.*, vol. 103, pp. 157–165, 2015

32. L. H. Saw, A. A. O. Tay, and L. W. Zhang, "Thermal management of lithium-ion battery pack with liquid cooling," *Annu. IEEE Semicond. Therm. Meas. Manag. Symp. San Jose, CA, USA*, vol. 2015-April, no. March, pp. 298–302, 2015. 10.1109/SEMI-THERM.2015.7100176.

33. R. Parrish *et al.*, "Voltec battery design and manufacturing," *SAE 2011 World Congr. Exhib.*, 2011

34. Y. Huo, Z. Rao, X. Liu, and J. Zhao, "Investigation of power battery thermal management by using mini-channel cold plate," *Energy Convers. Manag.*, vol. 89, pp. 387–395, 2015

35. L. W. Jin, P. S. Lee, X. X. Kong, Y. Fan, and S. K. Chou, "Ultra-thin mini-channel LCP for EV battery thermal management," *Appl. Energy*, vol. 113, pp. 1786–1794, 2014

36. A. Jarrett and I. Y. Kim, "Design optimization of electric vehicle battery cooling plates for thermal performance," *J. Power Sources*, vol. 196, no. 23, pp. 10359–10368, 2011

37. M. R. Khan, M. J. Swierczynski, and S. K. Kær, "Towards an ultimate battery thermal management system: A review," *Batteries*, vol. 3, no. 1, 2017

38. T. Talluri, T. H. Kim, and K. J. Shin, "Analysis of a battery pack with a phase change material for the extreme temperature conditions of an electrical vehicle," *Energies*, vol. 13, no. 3, 2020

39. L. Zhu, R. F. Boehm, Y. Wang, C. Halford, and Y. Sun, "Water immersion cooling of PV cells in a high concentration system," *Sol. Energy Mater. Sol. Cells*, vol. 95, no. 2, pp. 538–545, 2011

NOMENCLATURE

LiB	Lithium-ion battery
BTMS	Battery thermal management system
BP	Battery pack
EV	Electric vehicle
HEV	Hybrid electric vehicle
PHEV	Plug-in hybrid vehicle
ICE	Internal combustion engine
PCM	Phase change material

Chapter 13

Performance Evaluation of Diesel Engine with Fuels Prepared from Hydrogen and Nanoparticle Blended Biodiesel by Varying Injection Pressure

Mohammad Ashad Ghani Nasim, Mohd Parvez, Osama Khan, Gulam Hasnain Warsi, Md Hassaan, and Ashok K. Dewangan

CONTENTS

13.1 Introduction 231
13.2 Experimental Apparatus and Procedure 233
    13.2.1 Blend Preparation 234
    13.2.2 Nanoparticle Mixing 234
    13.2.3 Experimental Set-Up 235
    13.2.4 Physio-Chemical Characterization of Diesel Blends 235
    13.2.5 Response Surface Methodology (RSM) 236
    13.2.6 Artificial Neural Fuzzy Interface System (ANFIS) 237
13.3 Results and Discussion 240
    13.3.1 Brake Thermal Efficiency 240
    13.3.2 CO Emission 241
    13.3.3 UBHC Emissions 242
    13.3.4 NOx Emission 243
13.4 Conclusions 244
References 245

13.1 INTRODUCTION

Conventional engines used in automotive applications have several advantages evident from the past established with respect to energy efficiency, reliability and robustness. Conversely, these engines are also prime producers of dangerous exhaust emissions which impact the environment dangerous to human beings. Since the detection of the COVID-19 virus, scientists have been alerted and alarmed by the interconnection of engine exhaust and number of cases. Since these viruses have a strong affinity for pollution particles binding with them and increasing the risk of higher infections being detected at a large scale. Due to its ability to reduce exhaust emissions, a biodiesel fuel has been

DOI: 10.1201/9781003395768-13

232 Thermal Energy Systems

used in engine which has been used in several studies to resolve such a problem [1–3]. Furthermore, the inherent rise in gasoline prices and limited supply has further tightened the screws and motivated researchers to opt for clean and renewable fuel. In this perspective, hydrogen blending in biodiesel is regarded as an attractive proposition since zero-carbon fuel facilitates the creation of only water vapour and trace levels of NOx.

Previous studies have highlighted the importance of introducing hydrogen in biodiesel which can create a fuel that can be utilized in internal combustion engines with better prospects for the future [4–17]. Previous literature has also established the use of nanoparticles to reduce NOx emissions by about 75%. NOx emissions are reduced as the hydrogen proportion was raised, conversely increased with compression ratio (CR). Another advantage of adding hydrogen and nanoparticles is the reduction in fuel consumption levels since it facilitates a much cleaner combustion process with a stable flame front. When compared to pure biodiesel, hydrogen and NP-blended biofuel facilitate increased heat release tare (HRR) and pressure peak due to a shorter ignition delay (ID). One study discovered that using hydrogen at low and medium loads in a diesel engine with higher NP concentration yields favourable engine performance, thereby justifying its prospects for diesel engine usage.

This chapter explores the optimum concentrations of NPs and hydrogen coupled with proper operating conditions. One such input operating parameter is the ignition pressure, where a rise in the value is detected in final pressure with an increase in blending hydrogen percentage. Also, for loads larger than 80%, there was a pressure drop while the HRR was reported to be lower for both fuel blends. Another prospect detected by adding hydrogen blends was seen with NO being transformed to NO_2 when hydrogen was injected into the intake manifold of the engine [6]. This occurs as a result of the post-combustion process. The addition of NPs also significantly reduces the NOx emission levels, preferably due to proper combustion conditions and increased surface area. In addition, when compared to fossil fuels, the use of NPs and hydrogen in diesel engines lowered pollutants, notably CO [7]. In comparison to lesser loads, the combustion patterns changed dramatically, giving increased HRR and peak pressure. For all loading circumstances, the emission and performance characteristics remained identical with a 5% hydrogen supplementation. However, increasing the amount of hydrogen improved the engine's performance and reduced emissions effectively.

Variation in ignition pressure and compression ratio (CR) is a prime factor in engines blended with biodiesel, NPs and hydrogen while estimating performance and emission parameters. NOx, CO and HC emissions were also considerably altered while varying above operating conditions. Because of the reduced biodiesel quantity, there was an increase in HRR and BTE. By increasing the gasoline proportion, there was a considerable shift in the lowering of BSFC. Multiple studies used a lowered compression ratio to perform dual-fuel combustion using a diesel oxidation catalyst. The engine performed better and produced fewer emissions at lower CRs, according to the findings.

The engine performance and emission characteristics can be enhanced by applying alternative fuels with optimizing the operating parameters. Various techniques have been employed to optimize emission and performance parameters [18–23]. Response surface methodology (RSM) is an efficient technique to overcome the complexity observed during the experimentations [24,25]. Khan et al. [26] also optimized the parameters with different techniques (such as Taguchi and LSSVM). RSM also takes the least time to reduce the number of trials performs for each experiment [27]. Many studies have been reported by researchers [28] which focus on the efficacy of the RSM optimization technique for the input process parameters utilized in engine design.

Hosoz et al. [21] created computational fluid dynamics (CFD) models and used test data to validate them. The results revealed that increasing the injection angle resulted in the highest in-cylinder after TDC, and that BTE decreased with lower APID and vice versa. On a six-cylinder turbocharged intercooler DF, Mostafaei et al. [22] examined a variation of natural gas flow rate from 0% to 75% at a fixed air and diesel consumption rate. The tests were carried out at a speed of 1350 rpm with a load of 25%. The HRR curve was discovered to shift from a single peak to two peaks. The combustion was stated to be similar to that of an HCCI engine operating at 10% load. Kumar et al. [23] investigated the diesel/ethyl glycol (EG) dual fuel utilizing four different ethyl glycol energy ratios, EG0, EG10, and EG15, respectively. The cylinder peak pressure and HRR increased when EG was used in fumigation mode. In addition, the engine's BTE increased as a result of the shorter combustion time (CD). There was also a small increase in BSFC. H2 has a greater auto-critical temperature, allowing it to operate at a higher CR [24]. At 75% load, the DF engine with H2 produced improved BTE with reduced smoke and CO, but NOx emissions were unaffected [25,26].

Regardless of the potential advantage of combined hybrid fuel of NPs and hydrogen in diesel engines, a comprehensive review of the literature disclosed that the use of the hydrogen and biosynthesized cerium oxide amalgamation with Water hyacinth biodiesel in diesel engine has not been tested for varying compression ratios (CR), varying NP levels and hydrogen fuel blending concentration (HBC). As a result, the performance of a diesel engine using amalgamated blend fuel combinations is evaluated at various operating conditions such as CR, load and IP in this experimental analysis.

13.2 EXPERIMENTAL APPARATUS AND PROCEDURE

This research has been classified into four segments. Its first segment discusses the many types of energies and related operational characteristics. The experimental setup for investigating engine performance, acoustics and emission characteristics is detailed in the next segment. The science underlying the use of Response Surface Methodology and Artificial Neural fuzzy interface system is explained in the 3rd and 4th segments of this research, respectively.

13.2.1 Blend Preparation

The procurement process for Water Hyacinth oil and the preparation of various blends with hydrogen in detail has been discussed [26]. The raw Water Hyacinth oil is procured from a pond in Okhla Jamia Nagar, New Delhi as shown in Figure 13.1. A transesterification process is used to prepare Biodiesel amalgamated with hydrogen. The different blends of fuels were prepared and their properties will have an impact on engine characteristics (performance, combustion and emissions).

13.2.2 Nanoparticle Mixing

As discussed above, the shortcomings of biodiesel application in a diesel engine can be nullified by mixing nano-additives with biodiesel-diesel blends which provide superior performance parameters. It has already been established through a literature survey that mixing nanoparticles enables a larger surface area with a potential drop in viscosity and density levels. Normally nano-additives and biodiesel cannot be mixed directly with each other. It requires a chemical catalyst and an energy-imparting process for effective mixing. This is furnished by mixing surfactant 30 as a catalyst while ultrasonic horn provides the necessary energy addition to the reaction for boosting the inter-mixing capability. The nanofluids with metal particles are easily miscible with biodiesels. Initially, the nano-additives are weighed and combined with normal water to form a nano-fluid. The mixture is then positioned under the ultrasonic reactor at

Figure 13.1 (a) *Eichhornia crassipes* site and (b) *Eichhornia crassipes* stems.

90–100 kHz for 20 minutes. The size of nanoparticles applied in this investigation is approximately 30 nm.

13.2.3 Experimental Set-Up

The engine performance, emission and combustion characteristics are performed [26] with the help of an experimental setup as shown in Figure 13.2. A common rail direct injection (CRDI) diesel stationary engine is used to investigate. The performance and emission parameters are predicted by altering the engine load depending on the speed of the engine. An eddy current dynamometer helps to measure the engine load. Its combination with the diesel engine was employed to regulate the engine torque. K-type thermocouple was used to measure the temperature of exhaust gas. The technical specification of a diesel engine is given in Table 13.1. A separate panel box, i.e., fuel tank, air box, transmitters for air and fuel flow was embedded with the experimental setup. Gas analyser is used to measure the engine emission exhausts gases.

13.2.4 Physio-Chemical Characterization of Diesel Blends

The concentrations of nanoparticles (Petro-diesel and cerium oxide) with fuel affect the fuel properties to observe the feasibility of fuel in CI engines. The prepared blends were compared with some standard and their physiochemical properties are shown in Table 13.2. A capillary viscometer is used to measure the viscosity of the fuel blends. The density and calorific values were obtained using a hydrometer and a bomb calorimeter. The pressure of the barrel was monitored using a piezoelectric pressure sensor. Applying the

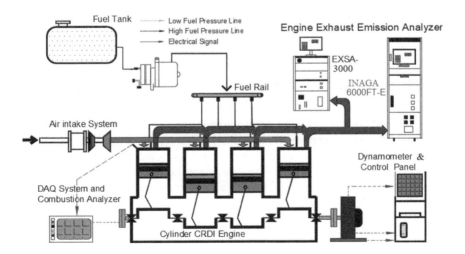

Figure 13.2 Experimental setup of PETTER-AV engine.

236 Thermal Energy Systems

Table 13.1 Technical details of the engine

S. no.	Component	Specification
1	Engine make	PETTER -AV1
2	Engine type	CRDI, 4-stroke, DI, water-cooled
3	Bore	80 mm
4	Stroke	110 mm
5	Rated power	5 BHP at 1500 rpm
7	Compression ratio	19:1
8	Dynamometer	Eddy current
10	Load sensor	Load cell, type strain gauge
11	Compression pressure	48.15 kg/cm^2

Table 13.2 Physiochemical properties of test fuels with ASTM standard

Properties	B30 (ECO)+cerium oxide	B30 (ECO)	Diesel	ASTM limit
Density at 15°C (kg/m^3)	886	908	841	860–900
Kinematic viscosity (cSt)	3.50	2.9	4.56	2.52–7.5
Calorific value (MJ/kg)	48.8	41.74	44.85	Min. 33
Flash point (°C)	71.66	104	51	Min. 130
FFA (%)	–	1.14	0.0014	Max. 2
Cetane number (°C)	58	57	51.3	Min. 45

fundamental rule of thermodynamic study [27,28] and determined HRR for every CA for every 50 cycles. The average of five readings was used for the result analysis. The compression ratio varied between 17 and 19, and the hydrogen fuel flow rate varied from 0 to 1.2 kg/h, respectively. All in all, the effects of load, CR and IP on diesel engine performance were investigated.

13.2.5 Response Surface Methodology (RSM)

To identify the correlation and establish related equations, RSM employs a range of well-conducted tests. It's important to remember that this is only an approximation statistical model. However, it is a tried-and-true approach that has been utilized by countless researchers to enhance a product, service or process. The test data is optimized by employing response surface regression [26,28], and the polynomials model of the second-order determined by the equation (13.1).

$$Y = \beta_o + \sum_{i=1}^{k} \beta_i X_i + \sum_{i=1}^{k} \beta_{ii} X_i^2 + \sum_{j \geq i}^{k} \beta_{ij} X_i X_j + \varepsilon \dots\dots\dots\dots\dots\dots\dots\dots\dots \quad (13.1)$$

RSM's CCRD methodology is used to obtain the desirable blends of biofuel at different operating conditions with nanoparticle concentration. The outcomes obtained from the advisability methodology are verified and validated with other researchers' results. This study aims to reduce exhaust gas pollutants and engine vibration and sound while concurrently maximizing engine performance.

RSM has been used to model nonlinear connections among both input and output variables in this research. RSM has the benefit of being able to refine alternatives with decimals of factor levels with factorial architecture. Using experimental data from orthogonal arrays, RSM models statistical connections amongst inputs and responses. To identify the problem, a linear or second-order polynomial model is usually built to get the correlation between the input and response variables. When using the response surface approach, the first step is to create a correlated equation function among the input and the response; if a low-order polynomial function could properly forecast the connection, the first-order or linear model will be the estimation function.

Engine performance and emissions were formerly expressed in terms of brake thermal efficiency, specific fuel consumption, smoke, NOx emissions and so on. Normally, these factors are taken into consideration as a response. It is vital to get more than one response available. These response variables are critical for obtaining useful information to evaluate engine performance and emissions characteristics. Based on the impacts of five input factors (load) from varied CR and IP rates on six engine output responses, two-dimensional contour plots and their related 3D interactive surface plots were given in this work (brake thermal efficiency, BSFC, HC, CO, NOx, vibrations and sound). The final developed table is displayed in Table 13.4 as per the experimental parameters (Table 13.3).

13.2.6 Artificial Neural Fuzzy Interface System (ANFIS)

Artificial neural network (ANN) is (also known as neural network which is inspired by biological neural network) a branch of artificial intelligence (AI)

Table 13.3 Experimental parameters and their ranges

S. no.	Input parameters	Coded value	Low	High	Range
1	Load (%)	2	20	100	20, 40, 60, 80, 100
2	Hydrogen blend (%)	1	0 or 0%	1.2 or 40%	0 (0%), 0.3 (10%), 0.6 (20%), 0.9 (30%), 1.2(40%)
3	Nanoparticle concentration (NPC)	0	0	80	0, 20, 40, 60, 80
4	Injection pressure, IP (bar)	−1	180	220	180, 190, 200, 210, 220
5	Compression ratio, CR	−2	17	19	17, 17.5, 18, 18.5, 19

Table 13.4 Experimental results of diesel engine

| S.no | Input criteria | | | | | Outcomes | | | | |
| | | | | | | Performance | | Emissions | | |
	Load (%)	Hydrogen blend (%)	IP (bar)	NPC	CR	BTE	BSEC	NO_x	UBHC	CO
1	80	40	190	80	17.5	35.81	0.29	721.38	34.74	20.58
2	80	10	190	40	18.5	30.87	0.36	544.29	39.20	20.02
3	80	10	210	60	17.5	30.69	0.37	735.69	39.66	17.06
4	80	10	210	80	18.5	30.64	0.37	743.93	39.83	19.69
5	80	30	190	0	18.5	35.48	0.29	732.35	34.95	17.22
6	80	10	190	20	17.5	29.65	0.37	825.82	41.57	18.06
7	80	30	210	40	17.5	34.27	0.31	768.80	36.60	17.55
8	60	20	200	80	17	30.35	0.35	670.47	38.84	21.52
9	60	20	200	0	18	30.74	0.33	646.49	37.14	21.07
10	60	0	200	20	18	29.08	0.36	637.01	39.77	21.41
11	60	20	200	40	18	30.73	0.35	619.92	37.11	21.48
12	80	30	210	80	18.5	36.48	0.28	643.94	33.28	21.34
13	60	40	200	20	18	30.73	0.35	677.22	36.44	21.13
14	60	20	200	40	18	30.35	0.37	659.81	38.31	21.16
15	60	20	200	60	18	27.63	0.39	748.98	42.26	21.86
16	60	20	200	80	19	34.09	0.32	588.73	36.12	20.67
17	60	20	200	0	18	30.74	0.36	627.03	37.13	20.98
18	60	20	200	0	18	34.74	0.31	570.65	33.13	24.02
19	60	20	220	80	18	33.25	0.32	607.31	36.59	21.13

20	60	20	180	40	18	32.51	0.33	637.68	39.24	21.76
21	60	20	200	60	18	30.74	0.36	644.08	37.12	21.03
22	60	20	200	80	18	30.74	0.36	637.68	37.12	20.91
23	40	30	210	20	18.5	28.64	0.38	747.45	39.61	20.46
24	40	10	210	40	17.5	24.91	0.43	672.99	40.09	29.82
25	40	10	190	40	17.5	26.21	0.40	614.13	38.13	29.89
26	40	30	210	60	17.5	27.69	0.40	789.28	40.26	21.89
27	40	30	190	20	17.5	27.04	0.39	858.89	41.42	17.05
28	40	10	190	40	18.5	26.46	0.41	704.11	39.75	20.58
29	40	10	210	60	18.5	27.99	0.39	719.78	39.22	19.69
30	40	30	190	20	18.5	27.96	0.39	729.81	39.56	20.02
31	100	20	200	0	18	30.50	0.37	822.41	38.15	22.11
32	20	20	200	80	18	25.81	0.40	715.12	43.24	24.37

which is taught to identify a relationship in inputs and the expected results after training by the train data. It's a type of AI that's meant to mimic the functioning of the animal brain cell. Their brain is made up of a collection of biological units that are linked together and transfer the signal to their corresponding biological units. The units are known as "brain cells" or "neurons." Neural networks, like animal brain cells, are made up of structural components called neurons. The layers are made up of extremely basic computer components, and the way they interact affects the network's performance. Every layer's neuron output is weighted using nonlinear activation function before being used as the input for the following layers.

13.3 RESULTS AND DISCUSSION

13.3.1 Brake Thermal Efficiency

The results at five different input conditions in a diesel engine are highlighted in this section. The rate of hydrogen flow was also varied to understand the combined action of NPs and hydrogen induction. The function of braking power in an engine to the heat input from the fuel is known as brake thermal efficiency. Figure 13.3 shows the relationship between BTE and operating parameters. It is apparent that for all three fuels injected at all loads, BTEs rise with CR.

This might be attributed to improve combustion as a result of greater compression ratio and higher cylinder temperature. Infusions of higher amount of WH biodiesel (WHB) significantly decreased BTE primarily because of its superior combustion characteristics, diesel fuel surpasses biodiesels at all loads. When comparing a conventional diesel engine and a hybrid NP-hydrogen blended CI engine, the hybrid engine-fueled brake

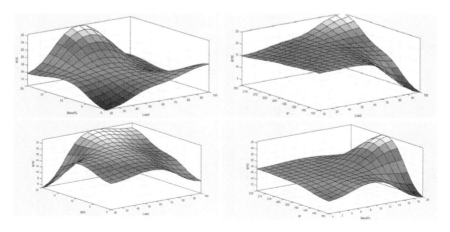

Figure 13.3 Brake thermal efficiency vs. the operating parameters.

thermal efficiency increased by 20–22%, when compared to a normal diesel-fueled engine. Due to its lowered viscosity due to the inclusion of cerium oxide nanoparticles, 80 ppm concentration depicted the greatest BTE which can lead to upgradation in fuel atomization. This is further coupled with appreciable hydrogen content which improves the existing efficiency due to superior combustion behaviour producing a steadier flame front. The higher calorific value (45 MJ/kg) of diesel fuel compared to existing blends of WHB (40.72 MJ/kg) is countered and nullified by additions of hydrogen and cerium oxide NPs, thereby contributing to its higher BTE. It is well understood that greater calorific fuels require less energy to produce the same amount of power. As a result of a combination of its higher calorific value and lower fuel usage, the BTE of hybrid-prepared fuel was comparatively higher.

The greatest BTE is 33.12% and is achieved when hybrid NP-hydrogen fuel is fed at 80 ppm and 40% hydrogen content in a diesel engine. Yet, for both workloads, an increase in hydrogen flow rate was observed to be associated with an increase in BTE value. It may be due to a higher effective burning mechanism and a quicker hydrogen flame. Hydrogen-CO-WHB activity had a greater BTE than H2-WHB activity due to the better combustion characteristics achieved by surplus surface area of cerium oxide nanoparticles. The engine displayed a 12–14% reduction in BTE value with only biodiesel relative to CI operation at the maximum attainable HFR of 0.24 kg/h and knock-free operation at 80% load.

13.3.2 CO Emission

The particle of solid soot embedded in the exhaust gas is the prime cause of smoke to be produced. In general, Figure 13.4 demonstrates that when the load increases, smoke emissions for all of the tested fuels progressively rise. The rising compression ratio is also associated with a constant decrease in smoke emissions. Lower smoke emissions from high compression ratios can be ascribed to higher air density and in-cylinder temperature, which results in improved fuel-air mixture, which promotes wholesome combustion, thereby improving soot oxidation, and simultaneously reducing smoke emissions.

The zenith smoke emissions are consistently produced in blends with high content of biodiesel and lower content of NPs and hydrogen. Also, engines operated at lower compression ratios produce higher smoke in diesel engines as compared to higher CR. Moreover, as compared to the Hydrogen-Cerium oxide-biodiesel combination, Hydrogen-biodiesel produced more smoke. At a CR of 17.5 and 100% load, the smoke emissions of a CI engine with H2-cerium oxide-biodiesel were nearly identical. Previous research [30,31] came up with similar outcomes. Figure 13.4 shows the variance in pollutant emissions in DF mode when the HFR is adjusted to 80% and 10%. Hybrid fuel injected at 190 bar with 100% engine load in CI mode

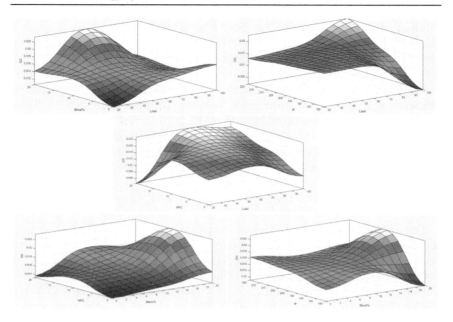

Figure 13.4 Surface plots for CO vs. input conditions.

produces the maximum smoke emission of 66 HSU. Diesel fuel supplied at 220 bar and 60% engine load, on the other hand, emits the least amount of smog at 40 HSU. The volume of oxygen required for full-burning fuel was insufficient as a result of hydrogen induction. Evidently, hydrogen-cerium oxide-biodiesel mix created less smog than hydrogen-biodiesel, given the lesser viscosity and cetane number of biodiesels. At an HFR of 1.4 kg/h at 60% load, the hybrid-fueled engine produced lowered smoke emissions as compared to other blends.

13.3.3 UBHC Emissions

Figures 13.5 show the variation in unburnt hydrocarbons (UBHC) for various blends of hydrogen and NP at different levels and operating conditions. Overall, UBHC emissions for all the tested fuels progressively rise with increasing load. With the subsequent rise in the CR value, there is a consistent drop in UBHC emissions. Lower UBHC emissions from high compression ratios might be explained by a better mixing process between the air and the fuel as a result of a greater in-cylinder temperature during the compression stroke as a result of increased CR. This would allow for a more thorough combustion, improve the post-exudation hydrocarbon, and aid CO_2 conversion. Conventional biodiesel-hydrogen consistently produces the greatest HC emissions at lowered CR values, whereas hydrogen-cerium oxide-biodiesel fuel consistently produces the lowest at higher CR value.

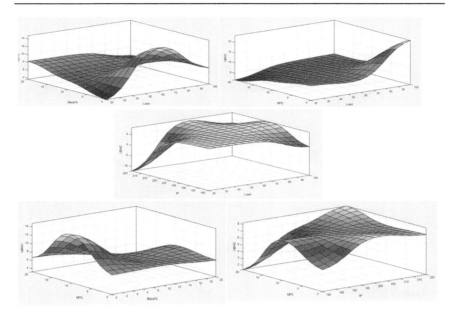

Figure 13.5 Surface plots for UBHC vs. input conditions (a–e).

In CI mode, diesel fuel fed at an IP of 190 bar and 100% engine load produces the maximum UBHC exhaust at. On the other hand, at hydrogen-cerium oxide-diesel fuel fed at 220 bar and 60% engine load emits the least amount of UBHC, as demonstrated in Figures 13.5(a-e), HC and CO have a similar trend. With a rise in hydrogen flow rate, engine produces reduced UBHC emissions. This might be attributed to better mixture combustion as a result of higher cylinder pressure and heat release rate. The hydrogen fuel has a quicker burning rate that quickly approaches the cylinder wall, leading in full burning and reduced UBHC levels. At the maximum feasible hydrogen flow rate of 1.4 kg/h at 60% load, the engine demonstrated around 8–14% lower UBHC using a hybrid combination of hydrogen-cerium oxide biofuel blends.

13.3.4 NOx Emission

Figure 13.6(a-f) shows the NOx emission with different blends of fuels. Nitrogen oxides (NOx) are a mixture of oxygen and nitrogen gas molecules. They are mostly made up of nitric oxide (NO) and nitrogen dioxide (NO_2). NOx production is influenced by oxygen levels, combustion chamber temperature and reaction residence time. The difference in NOx generated by the engine at different HFRs of 80 and 100%. It indicates that diesel fuel injected at 27 bTDC and full engine load in DF mode five emits the most NOx (1400 ppm). Diesel fuel injected at 23 bTDC and 80% engine load in CI mode, on the other hand, emits the least amount of NOx (1095 ppm). The rise in HFR

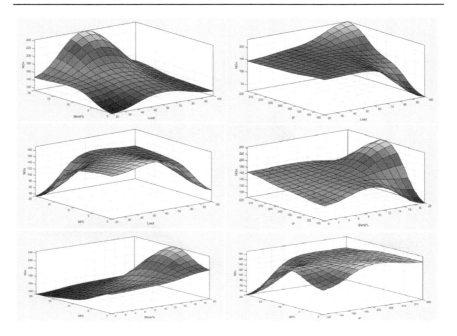

Figure 13.6 NOx emission rate with respect to operating conditions.

caused a rise in NOx emissions. The enhanced NOx emission might be due to the increased temperature of the burnt gases and greater pressure in the combustion chamber as a response of the rapid rate of consumption of H2 with higher HRR. With biodiesel and an HFR of 0.24 kg/h at 80% load, the DF engine produced approximately 11% to 22% more NOx.

Figure 13.6a depicts the impact of CR and BP on NOx emission fluctuations. Owing to greater heat emission rate and gas temperature prevailed due to higher flame velocity of H_2, an increase in NOx was noticed with CR at all loads DF engine operating. Biodiesel produced less NOx than diesel. At an HFR of 0.24 kg/h and at 80% load, DF engine NOx was 9 to 11% higher with biodiesels than with conventional engines. The literature [30] revealed similar conclusions. Diesel fuel, with a CR of 17.5, regularly produces the most NOx emissions, whereas BUO, with a CR of 15.5, usually produces the lowest. The lower compression ratio may have minimized in-cylinder temperatures, which might explain the reduced NOx emissions. As a result, at low compression ratios, flame temperatures would limit the production of NOx emissions.

13.4 CONCLUSIONS

The overall engine performance with different CR, HFR and CC forms is explored in this study. The following are the findings of the experiments:

- H_2 fuel induction saves liquid fuel. Except for NOx, DF engine running with a CR of 17.5, HFR of 0.24 kg/h and TRCC shape resulted in improved performance and reduced emissions.
- When compared to the CI mod, this engine produced roughly 8–12% lower BTE and 5–9% reduced engine smoke at a CR of 17.5, HFR of 0.24 kg/h and a TRCC shape.
- At the optimum condition of 80% load, there was a 3–6% reduced HC and a 20–27% lower CO.
- When compared to the CI mode, the DF mode produced around 20–29% higher NOx emissions at 80% load.
- In DF operation at 80% load, both PP and HRR were 12–15% greater than in CI mode.
- The RSM and ANN models' outputs were found to be in good agreement with experimental data, with ratios ranging from 90% to 93.5%.
- Overall, the DF engine operation with both biodiesels and H_2 was smooth. In addition to addressing the country's energy security, DF engine running with these fuel combinations provides complete freedom from fossil diesel fuel.

REFERENCES

1. Parvez, M, Khan, ME, Khalid, F, Khan, O, Akram, W. (2021). A novel energy and exergy assessments of solar operated combined power and absorption refrigeration cogeneration cycle. In Patel, N., Bhoi, A. K., Padmanaban, S, Holm-Nielsen, J.B. (eds). *Electric Vehicles. Green Energy and Technology.* Singapore: Springer. 10.1007/978-981-15-9251-5_13
2. Nithya, S, Manigandan, S, Gunasekar, P, Devipriya, J, Saravanan, WS. (2018). The effect of engine emission on canola biodiesel blends with TiO_2. *Int J Ambient Energy*, 10, 1–4.
3. Naddaf, A, Heris, SZ. (2018). Experimental study on thermal conductivity and electrical conductivity of diesel oil-based nanofluids of graphene nanoplatelets and carbon nanotubes. *Int Commun Heat Mass Transfer*, 1(95), 116–122.
4. Meraj, M, Khan, ME, Tiwari, GN, Khan, O. (2019). Optimization of electrical power of solar cell of photovoltaic module for a given peak power and photovoltaic module area. In: Saha, P, Subbarao, P., and Sikarwar, B. (eds). *Advances in Fluid and Thermal Engineering. Lecture Notes in Mechanical Engineering.* Singapore: Springer. 10.1007/978-981-13-6416-7_40
5. Khan, S, Tomar, S, Fatima, M, Khan, MZ. (2022). Impact of artificial intelligent and industry 4.0 based products on consumer behaviour characteristics: A meta-analysis-based review. *Sustain Oper Comput*, 3, 218–225.
6. Fatima, M, Sherwani, NUK, Khan, S, Khan, MZ. (2022). Assessing and predicting operation variables for doctors employing industry 4.0 in health care industry using an adaptive neuro-fuzzy inference system (ANFIS) approach. *Sustain Oper Comput*, 3, 286–295.

7. Drakaki, E, Dessinioti, C, Antoniou, CV. (2014). Air pollution and the skin. Front Environ Sci, 2, 1–6. Health problems Central Pollution Control Board (CPCB).
8. Khan, O, Khan, MZ, Khan, ME, Goyal, A, Bhatt, BK, Khan, A, Parvez, M. (2021). Experimental analysis of solar powered disinfection tunnel mist spray system for coronavirus prevention in public and remote places. *Mater Today: Proceedings*, 46 (15), 6852–6858.
9. Parvez, M, Khalid, F, Khan, O. (2020). Thermodynamic performance assessment of solar-based combined power and absorption refrigeration cycle. *Int J Exergy*, 31(3), 232–248.
10. Seraj, MM, Khan, O, Khan, MZ, Parvez, M, Bhatt, BK, Ullah, A, Alam, MT. (2022). Analytical research of artificial intelligent models for machining industry under varying environmental strategies: An industry 4.0 approach. *Sustain Oper Comput*, 3, 176–187.
11. Parvez, M, Khan, O. (2020). Parametric simulation of biomass integrated gasification combined cycle (BIGCC) power plant using three different biomass materials. *Biomass Conver Biorefin*, 10 (4), 803–812.
12. Yadav, AK, Khan, O, Khan, ME. (2018). Utilization of high FFA landfill waste (leachates) as a feedstock for sustainable biodiesel production: Its characterization and engine performance evaluation. *Env Sci Pollut Res*, 10. 1007/s11356-018-3199-0
13. Modak, NM, Ghosh, DK, Panda, S, Sana, SS. (2018). Managing greenhouse gas emission cost and pricing policies in a two-echelon supply chain. *CIRP J Manufact Sci Technol*, 20, 1–11.
14. Khan, O, Khan, MZ, Khan, E, Bhatt, BK, Afzal, A, Ağbulut, U, Shaik, S. (2022). An enhancement in diesel engine performance, combustion, and emission attributes fueled with *Eichhornia crassipes* oil and copper oxide nanoparticles at different injection pressures. *Energy Sources, Part A: Recov Utilization Environ Effects*, 44(3), 6501–6522.
15. Simsek, S, Uslu, S. (2020). Investigation of the effects of biodiesel/2-ethylhexyl nitrate (EHN) fuel blends on diesel engine performance and emissions by response surface methodology (RSM). *Fuel*, 275, 118005.
16. Khan, O, Khan, ME, Parvez, M, Ahmed, KAAR, Ahmad, I. (2022). Extraction and Experimentation of Biodiesel Produced from Leachate Oils of Landfills Coupled with Nano-additives Aluminium Oxide and Copper Oxide on Diesel Engine. In: Khan, Z.H. (ed). *Nanomaterials for Innovative Energy Systems and Devices. Materials Horizons: From Nature to Nanomaterials*. Singapore: Springer.10.1007/978-981-19-0553-7_8
17. Gopal, K, Sathiyagnanam, AP, Rajesh Kumar, B, Saravanan, S, Rana, D, Sethuramasamyraja, B. (2018). Prediction of emissions and performance of a diesel engine fueled with n-octanol/diesel blends using response surface methodology. *J Cleaner Prod*, 184, 423–439.
18. Singh, NK, Singh, Y, Sharma, A, Kumar, S. (2020). Diesel engine performance and emission analysis running on jojoba biodiesel using intelligent hybrid prediction techniques. *Fuel*, 279, 118571.
19. Aghbashlo, M, Hosseinpour, S, Tabatabaei, M, Soufiyan, MM. (2019). Multi-objective exergetic and technical optimization of a piezoelectric ultrasonic reactor applied to synthesize biodiesel from waste cooking oil (WCO) using soft computing techniques. *Fuel*, 235, 100–112.

20. Odibi, C, Babaie, M, Zare, A, Nabi, MN, Bodisco, TA, Brown, RJ. Exergy analysis of a diesel engine with waste cooking biodiesel and triacetin. *Energy Conversion Manage*, 198, 111912.
21. Hosoz, M, Ertunc, HM, Karabektas, M, Ergen, G. (2013). ANFIS modelling of the performance and emissions of a diesel engine using diesel fuel and biodiesel blends. *Appl Thermal Engineer*, 60, 24–32.
22. Mostafaei, M. (2018). ANFIS models for prediction of biodiesel fuels cetane number using desirability function. *Fuel*, 216, 665–672.
23. Kumar, S. (2020). Estimation capabilities of biodiesel production from algae oil blend using adaptive neuro-fuzzy inference system (ANFIS). *Energy Sources, Part A: Recovery, Utilization Environmental Effects*, 42, 7.
24. Millo, F, Aryaa, P, Mallamo, F. (2018). Optimization of automotive diesel engine calibration using genetic algorithm techniques, *Energy*, 158, 807–819.
25. Khan, O, Khan, ME, Yadav, AK, Sharma, D. (2017). The ultrasonic-assisted optimization of biodiesel production from eucalyptus oil. *Energy Sources, Part A: Recovery, Utilization, Environment Effects*, 39(13), 1323–1331.
26. Khan, O, Khan, MZ, Bhatt, BK, Alam, MT, Tripathi, M. (2022). Multi-objective optimization of diesel engine performance, vibration and emission parameters employing blends of biodiesel, hydrogen and cerium oxide nano-particles with the aid of response surface methodology approach. *International J Hydrogen Energy*, ISSN 0360-3199. 10.1016/j.ijhydene.2022.04.044
27. Khan, O, Yadav, AK, Khan, ME, Parvez, M. (2019). Characterization of bioethanol obtained from *Eichhornia crassipes* plant; its emission and performance analysis on CI engine. *Energy Sources, Part A: Recovery, Utilization, Environment Effects*, 43, 1–11.
28. Khan, O, Khan, MZ, Ahmad, N, Qamer, A, Alam, MT, Siddiqui, AH. (2021). Performance and emission analysis on palm oil derived biodiesel coupled with aluminium oxide nanoparticles. *Materials Today: Proceedings*, 46, 6781-6786, ISSN 2214-7853. 10.1016/j.matpr.2021.04.338

Chapter 14

Effect of Antioxidant *Psidium guajava* Extract on the Stability of Oxidation of Various Biodiesels

Meetu Singh, Neerja, Amit Sarin, Deeptam Trivedi, Sujeet Kesharvani, Anjali Agrawal, and Gaurav Dwivedi

CONTENTS

14.1 Introduction 250
14.2 Stability of Oxidation 251
    14.2.1 Importance of Biodiesel Stability of Oxidation 251
    14.2.2 Characterization Methods for Stability of Oxidation...252
        14.2.2.1 Rancimat (EN-14112)................................252
    14.2.3 Fluorescence Spectroscopy 254
    14.2.4 Thermogravimetric Analysis and Differential Thermal Analysis (TGA/DTA)..255
    14.2.5 Light Reflectance Method (ASTM D-6468) 256
    14.2.6 Gravimetric Analysis 256
    14.2.7 Active Oxygen Method (AOM) 256
    14.2.8 Jet Fuel Thermal Oxidation Tester (JFTOT)................257
14.3 Fatty Acid Compositions for Various Fats and Oils 257
14.4 Impact of Oxidation of Biodiesel on Diesel Engines 258
14.5 Antioxidants...258
    14.5.1 Natural Antioxidant..259
    14.5.2 Synthetic Antioxidant...260
14.6 Materials and Methods 261
    14.6.1 Preparation of *Psidium guajava* Extract (PGE) 261
    14.6.2 Biodiesel Preparation..261
    14.6.3 Fatty Acid Methyl Ester (FAME) Proportion and Physicochemical Properties..261
    14.6.4 Identification of the Structural Composition of PGE ...262
    14.6.5 DPPH Scavenging Activity 262
    14.6.6 Total Phenolic Content 263
    14.6.7 Stability of Oxidation (OS) 263
14.7 Results and Discussion..263
    14.7.1 Physico-Chemical Properties of JBD 263
    14.7.2 Identification of functional groups in *Psidium guajava* 263
    14.7.3 DPPH Scavenging Effect 264

DOI: 10.1201/9781003395768-14

249

250 Thermal Energy Systems

14.7.4 Overall Phenolic Content (TPE) 267
14.7.5 Outcome of Guava Extract on the Stability
of Oxidation of JBD 267
14.8 Conclusion 269
References 270

14.1 INTRODUCTION

Energy is the basic unit of humanity and a critical aspect in every nation's progress. Diesel and petroleum-based fuels play a crucial role in the energy industry, also in the transportation sector, which contributes to the country's industrial growth. Due to a rapidly rising population and changing lifestyle, the globe has experienced an exponential surge in energy needs and resource problems over the last few decades. This increase in energy consumption, along with the depletion of fossil fuels supply, has prompted the scientific society to seek alternative fuels derived from inexhaustible energies that are ecologically benign, everlasting, and emit fewer greenhouse gases. Energy from biomass resources gained popularity due to its wide range of feedstocks, which largely consist of oils (triglycerides) from plants and animal fats, as well as growing attention and the need to depart from petrochemical-based technologies. Biodiesel is the most common renewable fuel on which much research is being conducted [1–3].

The recent decade has witnessed considerable advancements in this industry with varied biodiesel feedstock. Between many multiple feedstocks available, *Jatropha curcas* and *Pongamia pinnata* are the most researched plants [4–6], but it is necessary to discover more alternative feedstocks to ensure that a disciplined strategy could be proclaimed toward renewable resources, and this in mind, the research has been stretched with the experimenting on novel biodiesel from *Tectona grandis* oil [7–9]. Because these three oils are non-edible, there is no rivalry with food resources when they are used in the synthesis of biodiesels.

So far, most studies have used methanol as an alcohol in the transesterification procedure. Because methanol is sourced from fossil resources, biodiesel cannot be entirely described as renewable. Alcohols must be obtained from bio-based resources in order for biodiesel to be totally renewable. This class of alcohol includes ethanol, butanol and pentanol. The purpose of this research is to explore the effect of various higher alcohols on biodiesel production, yield and fuel quality of biodiesel.

The majority of second-generation oils are non-edible oils generated from a range of feedstocks, including lignocellulosic feedstocks and nonfood crops. Karanja, *Jatropha curcas*, *Polanga*, Linseed, *Moringa oleifera*, *Croton megalocarpus*, *Jojoba*, Chinese tallow and other oils fall within this category. In comparison to first-generation oils, the advantage of employing non-edible oils is that they do not place an undue load on food crops.

Since biodiesels were produced by transesterification of long-chain fatty acids from animal or vegetable fats, they were mono-alkyl esters. Biodiesel is ecologically friendly, however, it has drawbacks such as deteriorating stability of oxidation, cold flow properties, storage stability, and so forth. Cold flow qualities are factors that must be addressed while running compression-ignition engines in mild-temperature environments, such as the winter climate [10].

14.2 STABILITY OF OXIDATION

Stability of oxidation chiefly determines the resistance of biodiesel versus oxidation, which forecasts the quality and shelf life of any biodiesel. Due to the presence of various degrees of unsaturation in fatty acid chains, including allylic and *bis*-allylic sites, biodiesels are more sensitive to oxidative deterioration than petro-based diesel [11]. As can be seen from the following reaction process, oxidation happens throughout preservation as a sequence of reactions characterized as initiation, propagation and termination.

$$RH + I \xrightarrow[\text{Propagation}]{\text{Initiation}} R^\bullet + IH \quad ; \quad I \text{ is initiator radical}$$

$$R^\bullet + O_2 \to ROO^\bullet \quad \text{(Peroxide Radical)}$$

$$ROO^\bullet + RH \xrightarrow[\text{Termination}]{} ROOH + R^\bullet$$

$$R^\bullet + R^\bullet \to R - R$$

$$ROO^\bullet + ROO^\bullet \to \text{Product}$$

The stability of oxidation of biodiesels has been designated as a pronounced factor by numerous standards including EN-14214 and ASTM D-6751 method [12,13].

14.2.1 Importance of Biodiesel Stability of Oxidation

Fuel stability refers to a fuel's resistance to activities that can alter its properties and produce unwanted species. One or more of the following mechanisms might cause the characteristics of biodiesel fuel to deteriorate: (i) contact with oxygen in the surrounding air, which causes oxidation or autoxidation; (ii) excess heat-induced thermal or thermal-oxidative degradation; (iii) hydrolysis caused by contact with moisture or water in fuel lines and tanks; or (iv) the transfer of bacteria or fungi from water droplets or dust particles into the fuel, causing microbial contamination.

One of the most important properties of biodiesel during long-term storage is oxidation stability (OS). As a result, OS is a critical issue that biodiesel research must address. Oxidation produces oxidation products that may be damaging to fuel properties, change fuel characteristics and

252 Thermal Energy Systems

negatively affect the engine performance. Since oxidative deterioration of biodiesel, predominantly in the engine fuel systems, can cause issues with diesel engine performance and maintenance, specifications relating to OS should be American ASTM D-6751 and European EN-14214 standards.

Varnish deposits, insoluble sediments and acids can occur as a result of the oxidative breakdown of biodiesel. Fuel characteristics could change dramatically. Insoluble substances can block gasoline filters and lines in engines. The deterioration of engine parts and the formation of deposits by deposit-forming or corrosive species can impact engine performance. Biodiesel colour changes from yellow to brown as a result of oxidation, and it also develops polymers that can clog fuel filters and injectors. It also loses some of its heating value and gains more viscosity, acidity and cetane number.

Storage, thermal and oxidative stability are the three types of biodiesel stability. The synthesis of peroxides and hydroperoxides results in the formation of smaller molecules like ketones, alcohol, aldehydes and acid alcohols, according to OS. Thermal stability refers to biodiesel's capacity to endure the oxidation process at high temperatures, which produces a weight increase in both the biodiesel and the oil.

Natural antioxidants (in very modest quantities) found in biodiesel are damaged quicker at high temperatures, resulting in the noticeable oxidation process. Storage stability is concerned with the breakdown of biodiesel and its relationship with environmental variables such as contaminants, moisture, metal and light during storage.

14.2.2 Characterization Methods for Stability of Oxidation

Biodiesel OS may be assessed using a variety of methods, each of which gives information on a specific aspect of fuel stability. Some measures represent the oxidation propensity of the material, while others reveal the amount of oxidation products. The concentrations of oxidation components are tested to assess a fuel's relative resistance to oxidation. Approaches to determine oxidation stability can be classed based on the following criteria: initial fatty oil content, oxidation products, subsequent oxidation products, physical properties or any other factor indicating relative stability.

Specifications such as FAME content, linoleic acid content and iodine value with Z4 double bonds all help to reduce the material's oxidative sensitivity. To directly measure durability, the standardized test method EN-14112, an accelerated oxidation test that analyzes oxidation tendency, is utilized in conjunction with the Rancimat technology.

14.2.2.1 Rancimat (EN-14112)

The Rancimat technique is one of the most accurate methods for determining the oxidative stability of fat and oil. The American Oil Chemists Society (AOCS) uses this approach to test the stability of oxidation of fats

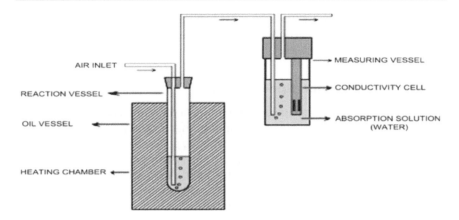

Figure 14.1 Schematic illustration of the Rancimat instrument's apparatus.

and oils. The Rancimat method is an accelerated oxidation test designed to determine the oxidative stability of FAME. The test is carried out at a high temperature, with the samples being exposed to an air stream to oxidize.

To begin the oxidation reactions, the sample is exposed to a higher temperature and then purge air through it. Primary oxidation results in the formation of peroxides. The peroxides are totally destroyed, yielding oxidation by-products such as organic acids with low molecular weight. Some gases are unintentionally released and flow into a container of filtered water. A conductivity measuring equipment is connected to the container. Water absorbs formic acids, acetic acids and other volatile gases produced during oxidation, enhancing conductivity. An oxidation curve is produced by systematic monitoring of this conductivity. The time at which the conductivity curve inverts is used to compute the induction period.

Because oxidation occurs progressively, the value of the conductivity curve grows very slowly at first. Oxidation accelerates as conductivity values rise, like an exponentially growing curve. The Rancimat equipment's system automatically collects the IP address. The impact of different antioxidants on enhancing the biodiesels' self-life may be tested utilizing this approach. The antioxidant's efficacy in avoiding oxidation of biodiesel may be measured based on the time of induction. The higher the induction period, the greater the antioxidant's capacity to increase the stability of oxidation of biodiesel.

A modified Rancimat experiment is carried out to verify the samples' storage stability. The samples are deposited in a heat exchanger during this operation. While the sample temperature approaches 80°C, air is expelled on its surface. The variations in conductivity are recorded by the conductivity cell, which is connected to the measurement vessel. For better outcomes, the same technique is performed for 6 hours at 200°C to examine the thermal stability of the sample. Because of its reliability and consistency,

254 Thermal Energy Systems

the modified Rancimat technique is widely employed. The main difficulty with this approach is avoiding adulteration of the sample with other volatile chemicals.

14.2.3 Fluorescence Spectroscopy

Fluorescence spectroscopy (also called fluorimetry or spectrofluorometry) is an electromagnetic spectroscopy technique that examines the fluorescence of a sample. It involves using a beam of light, generally ultraviolet light, to excite electrons in the molecules of certain compounds, causing them to emit light. Often, but not always, visible light. A complementary method is absorption spectroscopy. In the specific context of single-molecule fluorescence spectroscopy, changes in the intensity of light emitted from a single fluorophore or fluorophore pair are measured.

Fluorescence spectroscopy is another method for assessing the stability of FAME oxidation. The sample can be delivered to the fluorescence spectroscopic tests after being heated in the water bath. The number of oxidants and antioxidants in the sample is estimated by observing how their emission and excitation patterns vary. Fluorescence might be utilized to create novel methods and equipment for evaluating physicochemical changes in biodiesel after heat treatment. Fluorescence was used to efficiently evaluate the rate of oxidative degradation in rapeseed oil methyl ester (RME). To compare the Rancimat approach with fluorescence analysis, the presence of hydroperoxides and oligomers in rapeseed oil and its methyl esters was investigated using a fluorescence profile. The considerable growth in oligomer concentration was well recognized at a particular ageing time by using the excitation-emission matrix (EEM) model. At 6.8 h during the oxidation phase, the fluorescence signals for oligomers (440/505 nm) and hydroperoxides (400/450 nm) were dramatically raised. The results obtained with the Rancimat measurement were substantially equal. The observed induction period in the Rancimat test was 6.3 hours.

The fluorescence approach is a potent instrument for directly measuring oxidized products, whereas the Rancimat procedure indirectly evaluates oxidized products based on an increase in conductivity. Because of its sensitivity, precision and mobility, the fluorescence technique outperforms Rancimat. However, more testing is required to evaluate the effectiveness of the fluorescence approach with biodiesel derived from various sources.

The fluorescence data at various excitation wavelengths is recorded in EEMF. When charting fluorescence data versus excitation and emission wavelengths, a contour plot is created. Any fluorophore's fingerprint of fluorescence signals is determined by its concentration and excitation/ emission wavelengths. Because of its ease of use at the instrument level, EEMF spectroscopy is frequently employed. However, EEMF approaches have several disadvantages, such as the requirement for specialized skills for proper data processing.

14.2.4 Thermogravimetric Analysis and Differential Thermal Analysis (TGA/DTA)

TGA/DTA is a sensitive and precise technology for assessing the oxidative stability of biodiesel, according to various studies. A study looked at the relationship between the oil stability index (OSI) and the temperature of biodiesel. The temperature is used to measure fluctuations in any statistic's value. TGA is used to evaluate the thermo-oxidative and high-pressure oxidative properties of triglycerides. The unique profile of the crystallization oil is also documented. In TGA analysis, it is vital to monitor the sample's resistance to thermal-oxidative degradation. The chemical structure of a material determines its thermal stability. In general, saturated fatty acids are more thermally stable than unsaturated fatty acids.

In TGA, the sample is gradually heated across a temperature gradient, and the weight fluctuation of the material as a result of these parameters is observed. The weight change versus temperature plot graphically depicts the occurrence. This type of graph is known as a thermogram or thermogravimetric measurement. This approach provides information on the thermal degradation of a chemical to its volatile form. TGA can investigate either a pure component or a combination. Though biodiesel is a combination of mono-alkyl esters, most biodiesels degrade in a very restricted and specified temperature range. This is because of the distinction between biodiesel and its impurities.

As indicated in Figure 14.2, the sample is heated in a furnace. A thermocouple is a device that measures temperature changes. In general, an inert gas

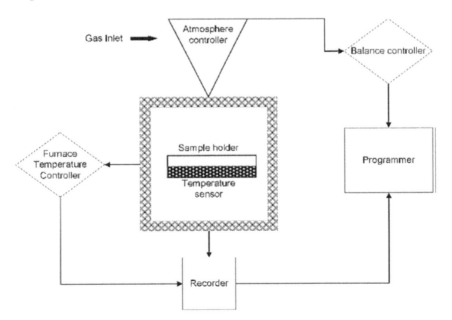

Figure 14.2 TGA analysis.

256   Thermal Energy Systems

is utilized to remove undesirable gases from the environment. To weigh something, a null point balancing is usually utilized. Whenever the weight changes, the balance will deviate significantly. The restoring force required to return to its former position will be precisely proportionate to the weight shift. The generated signal will be recorded. A contemporary thermo balance is temperature regulated and can operate in a broad range of cooling, heating and isothermal modes. The sample temperature variations may also be adjusted automatically. The temperature difference between sample and the relevant materials under similar heating circumstances is continually monitored as a function of the temperature (DTA). As a result, the system's heat absorbed or expelled is recorded. A differentiated thermal analysis depicts peaks and their accompanying temperatures. Peaks above zero show exothermic processes such as burning and adsorption, whereas those below zero indicate endothermic changes such as absorption, sublimation, desorption, vaporization and so on.

14.2.5 Light Reflectance Method (ASTM D-6468)

Reflectance is the effectiveness of a material's surface in reflecting radiant light. The energy created is represented by the proportion of incident electromagnetic power reflected at an interface. Reflectance spectra are graphs that show the relationship between reflectance and wavelength. The sample is heated to 150°C in open tubes with air contact for 90 or 180 minutes in this procedure. The sample has the required energy values after cooling and filtering. The colour of the filterable insoluble residue may influence reflectance test findings, resulting in measurement error. As a result, this method is not used for exact accuracy.

14.2.6 Gravimetric Analysis

The oxidative stability research employing gravimetric measurement of distillate fuel remains the best technique. Any biodiesel's insoluble sedimentation and polymerization properties may be determined by gravimetric analysis. The biodiesel sample is cooked for 16 hours at 95°C while air is filtered at 3 L/h. After thermal oxidation, the insoluble sediments are filtered. Sensitivity is achieved by fully dissolving the sediments in a tri-solvent mixture of toluene, acetone and methanol. The soluble polymers are precipitated with the help of iso-octane. The amount of insoluble residue produced in the test and control tests varies with oxidative stability. Gravimetric analysis, like light reflectance analysis, has limitations for fuels with higher residual oil content.

14.2.7 Active Oxygen Method (AOM)

The incumbent notion is also known as a PV approach since it is based on the measurement of peroxide value (PV). By heating a biodiesel sample at a constant temperature and purging air at a specified rate, the differential time duration for the time required to achieve the target peroxide is calculated. Because peroxides are fragile and degrade quickly, the repeatability of this technique is weak. As a result, AOM is regarded as the least effective approach in biofuel analysis. Furthermore, PV measuring becomes an unsuitable approach for monitoring the stability of oxidation. The AOM technique has limitations due to more tedious operations, high use of chlorinated solvents and variable results.

14.2.8 Jet Fuel Thermal Oxidation Tester (JFTOT)

JFTOT is essentially a standard approach for testing the thermal stability of oxidation of aviation gas turbine fuel at higher temperatures. The efficacy of the test fuel after exposure settings that mimic the gas turbine system is investigated here. The approach has the benefit of being able to measure the quantity of deposits that develop when liquid fuel comes into contact with a heated surface. In this process, the fuel is fed through a heater at a constant flow rate, and the degradation products contained in the fuel are captured by a stainless-steel filter.

The ellipsometry method is used to count the quantity of accumulated degradation products. The thickness of the film is determined visually using this approach. There is presently insufficient evidence to support this test, requiring more investigation.

14.3 FATTY ACID COMPOSITIONS FOR VARIOUS FATS AND OILS

In both human and animal diets, fats and oils are acknowledged as vital nutrients. They are the most concentrated source of energy of any meal, offer vital fatty acids (which are precursors for crucial hormones, the prostaglandins), considerably contribute to the sense of fullness after eating, are transporters for fat-soluble vitamins, and serve to make foods more appealing. Many foods contain various levels of fats and oils. Vegetable fats and oils, meats, dairy products, poultry, fish and nuts are the primary sources of fat in the diet. The majority of vegetables and fruits ingested in this manner contain just trace quantities of fat. The FA concentration of various oils and fats can vary greatly. FAME's FA composition is an important factor in oxidation.

Fatty acids are the building blocks of lipids and generally account for 90% of the fat in foods. This compound is of interest when reporting on the characterization of the lipid content of fats and oils. Saturated fatty acids – hydrocarbon chains with a single bond between each carbon atom – are predominantly found. Products of animal origin (meat, dairy) tend to have lower levels of high-density lipoprotein (HDL) cholesterol in the blood.

258 Thermal Energy Systems

Unsaturated fatty acids – represented by 1Double bonds in (singly unsaturated) or more (multiple unsaturated) carbon chains – commonly found for plants and seafood. There are few bonds because carbon atoms are connected by double bonds. It is available for hydrogen, so it has less hydrogen and becomes "unsaturated". *Cis* and *trans* – A term for the arrangement of chains of carbon atoms across a double bond. In the *cis*, in an array, the strands are on the same side of the double bond, resulting in a twisted shape. In the trans configuration, the strands are on opposite sides of the double bond and the strands are overall straight.

14.4 IMPACT OF OXIDATION OF BIODIESEL ON DIESEL ENGINES

There were constant sub-catastrophic difficulties in the few experiments recorded, characterized by enhanced accumulation on injectors and motor components, increased pressure drops through filters, and a few broken injectors and pumps. Another research revealed higher deposits on fuel pump components and corrosion in some fuel injector elements while using low-stability gasoline. After 1000 hours of operation on a B20 mix, a third analysis found severe loss of fuel injector performance and piston-ring damage with almost no fuel atomization, but no conclusions were reached. In other fuel injector tests, B20 created far more deposits than either straight biodiesel or clean petrol-diesel. It shows that when utilizing mixed fuels, additional issues may occur, which is consistent with previous research that indicated increased insoluble formation when combining biodiesel with Petro diesel.

Few writers have written about the effects of biodiesel oxidation on diesel engine equipment. Clearly establishing links between stability test data and real-world biodiesel engine performance remains a research challenge. It appears that agreement on the acceptable minimum biodiesel oxidation stability necessary to avoid engine problems has yet to be reached.

14.5 ANTIOXIDANTS

Antioxidant supplementation is an effective method of increasing oxidative stability processes and lowering Emissions of NOx from biodiesels. Antioxidants regulate and prevent the generation of by-products by preventing substrate auto-oxidation. Antioxidant identification entails making trade-offs between competing desired qualities. High solubility, effectiveness at small concentrations, extended shelf life, and negligible toxicity are a few of the most desirable features. Free radical terminators provide the capacity, also known as chain-breaking antioxidants. Hydroperoxide decomposers are the second antioxidants. Antioxidants are now being introduced to biodiesels

Effect of PGE on Various Biodiesels 259

to satisfy the mandated limitations. Antioxidants act as free radical scavengers, preventing free radical production in biodiesels throughout storage. Extremely labile hydrogen, which a peroxy radical may extract more easily than fatty oil or ester hydrogen, is one of the antioxidants [14]. The chemical reaction depicted below illustrates the mechanism used by all chain-breaking antioxidants.

$$ROO^{\bullet} + AH \rightarrow ROOH + A^{\bullet}$$
$$A^{\bullet} \rightarrow Stable$$

The functionality of numerous antioxidants used on distinct biodiesels to improve its stability of oxidation has already been discussed in the literature, with the majority of research using synthetic antioxidants but relatively few studies diagnosing the activity of natural antioxidants. However, because of its fossil-based origin, the use of synthetic antioxidants has been shown to damage the renewable quality of biodiesel while also raising manufacturing costs. Synthetic antioxidants, on the other hand, have been found to be hazardous due to its toxic character [15,16]. Thus, research for new trustworthy, benign nature-based chemicals with antioxidative action has been expanded in recent years to improve biodiesel stability of oxidation.

Antioxidants having low molecular weight have high antioxidant activity. Antioxidants with low molecular weight can easily pass through the biodiesel bulk. This permits the antioxidant to reach a greater number of inflexion points and limit radical formation. Although lower molecular weight antioxidants are more oxidatively stable, they are volatile and prone to evaporation. As a result, it may not be suitable for the long-term storage of biodiesel. Despite the fact that evaporation loss is significantly lower in biodiesel than it is in low molecular weight antioxidants, high molecular weight antioxidants are less likely to easily mobilize throughout the biodiesel. Low molecular weight, a low BDE value, and a high phenol value are among the selection criteria used to determine which antioxidants are good for biodiesel and how to prevent auto-oxidation based on their oxidation potentials.

14.5.1 Natural Antioxidant

Plant products are rich in phenols that can be utilized to augment antioxidants found in fats and oils. Natural antioxidants such as tocopherols and tocotrienols are abundant in vegetable oils. Plants provide significant promise for preventing fatty acid oxidation. Trans-esterification or refining methods can diminish or remove the bulk of these antioxidants. While biodiesels are made from unprocessed vegetable oils, they have more natural antioxidants and are more stable; yet, they lack fuel needs, limiting their practical usefulness. When 1000 ppm cashew nut shell liquid was added to beef tallow biodiesel, the IP increased by roughly 50%. The presence of cardanol and its

260 Thermal Energy Systems

derivatives in CNSL might explain the higher oxidative stability. When 1.5% leaf extract was applied, the IP rose from 5.0 to 14 hours, and the impact of *Pongamia pinnata* leaf extraction on the stability of oxidation of calophyllum biodiesel was superior. *Moringa oleifera* ethanolic leaf extract was determined to be the greatest antioxidant combination for biodiesel. Myricetin, a chemical discovered in *Moringa oleifera*, outperforms tocopherol and synthetic antioxidants in terms of antioxidant activity.

Natural antioxidants' influence on oxidative stability increases with alkyl chain length. Natural antioxidants at levels more than 1000 ppm in biodiesel having more than 80% unsaturated fatty acids boost induction duration enough to fulfil EN requirements. Despite the fact that the majority of natural antioxidants proved oxidative stability, their commercial relevance is unknown since they do not meet the EN criteria.

14.5.2 Synthetic Antioxidant

Synthetic phenolic antioxidants include butylated hydroxyanisole (BHA), BHT, propyl gallate (PG) and tertiary butylhydroquinone (TBHQ). Synthetic antioxidants have been shown to be more effective than natural antioxidants in improving oxidation stability. However, the bulk of these antioxidants' low biodegradability and toxicity prompted considerable concerns (Table 14.1).

Psidium guajava (commonly known as guava) is a substantial commercial fruit crop in tropical and subtropical nations, and its values claim supremacy over other fruits. Guava, like many other fruits and vegetables, contains a high concentration of antioxidant components such as polyphenols, ascorbic acid and carotenoids [17–20]. Many studies have been conducted on the antioxidant activity and therapeutic qualities of *Psidium*

Table 14.1 Various natural antioxidants' antioxidant efficacy in various biodiesel blends

S. no.	Biodiesel	Initial IP, h	IP after the addition of synthetic antioxidant at 1000 ppm				
			PY	TBHQ	PG	BHA	BHT
1.	Neem	1.39	44.5	2.15	30.2	4.54	2.41
2.	Poultry fat	0.67	13	4	5	12	6.5
3.	Cottonseed	6.57	26	30	21	9.2	9
4.	Waste cooking oil	5.9	31.95	29.44	29.9	13.8	10.84
5.	*Jatropha*	3.05	53.73	5.56	24.45	11.15	8.59
6.	Soybean	3.52	11.5	11	10	7	6
7.	*Pongamia*	1.82	34.35	6.19	0	5.02	4.88
8.	Sunflower seed	3.4	14	6.5	12	5.5	3.5
9.	Linseed	2.2	–	3	7	3	–
10.	Rapeseed	9.1	26.31	38.53	27.36	24.3	9.85

guajava, but no study has been done to investigate its potential as an antioxidant for any biodiesel.

The current study's main goal is to investigate and advance the stability of oxidation of biodiesels synthesized from non-edible oils viz. *Jatropha curcus*, *Pongamia pinnata* and *Tectona grandis* after blending with *PGE* as a substitute for synthetic antioxidants.

14.6 MATERIALS AND METHODS

Shrubberies of *Psidium guajava* were taken from the campus of IK Gujral Punjab Technical University in Punjab, India.

14.6.1 Preparation of *Psidium guajava* Extract (PGE)

Fresh leaves were washed with water, cut and sun-dried for three days before being dried in a hot air oven for three hours. The dry product was crushed into powder, and a sample of 10 g was carefully put into a glass container containing 200 mL of ethanol. The mixture was then mixed and centrifuged for 30 minutes at 3000 rpm. The extract was filtered via Whatman filter paper before being concentrated in a rotary evaporator at 40°C [21].

14.6.2 Biodiesel Preparation

The free fatty acid (FFA) content of the oils (*Jatropha*, *Pongamia* and *Tectona grandis*) was evaluated using acid number titration against potassium hydroxide with phenolphthalein as an indicator. Due to the low FFA content of biodiesel, the base-catalyzed transesterification process was used. In transesterification, methanol (1:6 M to oil) was individually added to the sodium methoxide as a catalyst (0.75 wt% oil) and agitated until the catalyst was entirely dissolved in methanol [22]. The stirred solution indicated above was introduced to the corresponding oils in the reactor, with the reaction temperature set to 65°C. The mixture was agitated at 400 rpm for one hour. Following the reaction's completion, the material was transferred to a separating funnel and left to settle overnight, resulting in the formation of two phases. The upper phase included methyl esters (biodiesel), whereas the bottom phase contained glycerin. The biodiesels were then washed four times with hot water to remove any remaining glycerin, unreacted catalyst, and soap from the transesterification process [23].

14.6.3 Fatty Acid Methyl Ester (FAME) Proportion and Physicochemical Properties

The quantities of methyl fatty acid ester (FAME) in samples were determined using a Nucon 5765 gas chromatograph with a microprocessor

262 Thermal Energy Systems

temperature controller and multi-ramp temperature programming. The biodiesel samples are injected into the apparatus, which then transfers the sample through a gas stream onto a column. A reference sample containing commons is also put into the device to evaluate the unidentified parts of the sample. The test sample's peak retention time and area are contrasted with those of the standard sample to determine concentration. The major physicochemical parameters of neat biodiesel samples, on the other hand, were determined using ASTM D-6751 and EN-14214 criteria [12,13].

14.6.4 Identification of the Structural Composition of PGE

PGE structural composition has been determined using the Fourier transform infrared (FTIR) spectrometry approach [24,25]. Using KBr pallets, the FTIR spectrum was obtained in the FTIR spectrophotometer (Perkin Elmer Frontier Instrument). The FTIR spectrum is commonly divided into two categories: functional group area and fingerprint region. The fingerprint region is 1450/cm to 4000/cm long, whereas the functional group region is 4000/cm long. Peaks in the functional group region are frequently associated with stretching vibrations of dominant functional groups. However, the fingerprint region's range is rather dubious for analysing precise groups because each molecule produces its own individual pattern of peaks [26].

14.6.5 DPPH Scavenging Activity

The DPPH free radical scavenging test focuses on electron transfer, which is a fast approach for assessing antioxidant activity using spectrophotometry. The free radical DPPH (2, 2-diphenyl-1-picryl-hydrazyl-hydrate) with an odd electron shows a maximum absorption at 517 nm due to its purple colour. When a DPPH solution is mixed with an antioxidant containing a labile hydrogen atom, DPPH is reduced to the pale-yellow non-radical Diphenylpicrylhydrazine (DPPHH) [27]. In summary, a 0.1 milli molar DPPH in ethanol solution was prepared, and 1 mL of this stock solution was added to 3 mL of different PGE concentrations in ethanol (50, 100, 500, 1000 and 2000 g per mL).

The following equation 14.1 was used to get the % DPPH scavenging effect:

$$\text{DPPH scavenging effect}(\%\ \text{inhibition}) = \frac{A0 - A1}{A0} \times 100 \qquad (14.1)$$

where A0 represented the absorbance without PGE and A1 represented the absorbance in the presence of the extracted sample.

Effect of PGE on Various Biodiesels   263

14.6.6 Total Phenolic Content

The Folin-Ciocalteau method, which relies on electron transfer from antioxidant compounds in an alkaline media, was used to determine the total phenolic content of PGE [28]. After 3 minutes, 2.5 ml of 7.5% NaHCO3 was added to 2.5 ml of prepared extract alcoholic solution and 2.5 ml of 10% Folin-reagent Ciocalteu's diluted in water to make the benchmark and blank samples. The materials were exposed to darkness for 40 minutes before being examined at 760 nm with a UV–vis–NIR 3600 spectrophotometer (SHIMADZU). The reaction was carried out with gallic acid, and the concentration of phenolic content in the antioxidant samples was determined as gallic acid equivalent (GAE) mg per 100 g using the calibration line.

14.6.7 Stability of Oxidation (OS)

PGE was added to biodiesel samples in 100 ppm increments from 100 to 1000 ppm. In line with specified testing parameters, the OS of biodiesel was evaluated using 893 pieces of specialized Biodiesel Rancimat apparatus (Metrohm, Switzerland). By passing a stream of air at a rate of 10 L/h through a 3 g biodiesel sample at a temperature of 110°C, the Rancimat approach promotes oxidation. The vapours produced during the oxidation reaction with air were introduced into a beaker containing 50 mL of demineralized water, and the flask is attached with an electrode coupled to a recording device that continuously analyzes the solution's conductivity, as shown in Figure 14.1. An oxidation graph is created in the recording equipment, the point of inflection of which is the induction period (IP), a measure of the sample's oxidation rate [29–37].

14.7 RESULTS AND DISCUSSION

14.7.1 Physico-Chemical Properties of JBD

Table 14.2 reviews the FAME alignment of JBD, PBD and TGBD samples and the physicochemical properties for the biodiesels are summarized in Table 14.3. The mean IP (induction period) for neat JBD, PBD and TGBD verified three times average to be 4.24, 3.71, and 3.43 h, respectively, that fulfil the specified limit of 3 h suggested by ASTM specifications but could not fulfil the EN-14112 specification of 6 h. To make the samples suitable as per standards, PGE as an antioxidant has been added to enhance their stability (Table 14.4).

14.7.2 Identification of functional groups in *Psidium guajava*

As depicted in Figure 14.2, the FTIR analysis of *Psidium guajava* indicates the frequency correlating to several functional groups. The peak at 3398 cm^{-1}

264 Thermal Energy Systems

Table 14.2 Specifications regarding the induction timeframe in different countries' biodiesel requirements

S. no.	Biodiesel standard	Limits (minimum induction time, hours)	Country/region
1	ASTM D-6751	3	USA
2	EN-14214	8	Europe
3	IS 15067	6	India
4	ANP 255	6	Brazil
5	SANS 1935	6	South Africa
6	Fuel Standard (Biodiesel), 2003	6	Australia
7	SNI 7182	8	Indonesia
8	GB/T 20828	6	China
9	B100 - FAME	6	Thailand

Table 14.3 FAME composition of biodiesels

FAME	Chemical formula	Content (wt %) for JBD	Content (wt %) for PBD	Content (wt %) for TGBD
Stearic	$C_{18}H_{36}O_2$	7.9	8.3	7.02
Arachidic	$C_{20}H_{40}O_2$	1.3	–	2.15
Oleic	$C_{18}H_{34}O_2$	42.7	71.5	42.71
Palmitic	$C_{16}H_{32}O_2$	5.3	9.6	13.10
Saturated	–	23.2	17.9	22.27
Linoleic	$C_{18}H_{32}O_2$	32.8	10.6	31.51
Unsaturated	–	76.8	82.1	74.69
Linolenic	$C_{18}H_{30}O_2$	–	–	0.47
Others	–	–	–	3.04

shows hydrogen bonding and correlates to the drawn-out vibration of a phenolic hydroxyl group (-OH). A signal at wavenumber 2812 cm^{-1} shows the existence of a (-CHO) group, whereas a peak at 1689 cm^{-1} designates the presence of a carbonyl group (C=O).

However, peaks at 1435, 963 and 771 cm^{-1} corresponding to fingerprint regions may be owing to the existence of -CH3 (methyl), R-O-R (ether), aromatic C-H bending, or phenol, respectively. *Psidium guajava*'s antioxidative efficacy is mostly due to the existence of phenolic groups in its structure.

14.7.3 DPPH Scavenging Effect

As seen in Figure 14.3, the suppression of DPPH radical is greatest whenever the amount of *Psidium guajava* extract (PGE) is held at 2000 g/mL,

Table 14.4 Physicochemical properties of JBD, PBD and TGBD along with proposed minimum limits

Property (units)	ASTM D-6751-08 test method	ASTM D 6751-08 limits	EN-14214 test Method	EN-14214 limits	Tectona grandis Biodiesel	Pongamia Biodiesel	Jatropha Biodiesel
Density (g/cc at 25°C)	D-1298	0.86–0.90	EN ISO 3675, EN ISO 12185	0.86–0.90	0.86	0.87	0.89
Cetane number	D-613	Min. 47	EN ISO 5165	Min. 51	54.3	49.6	53.4
Stability of oxidation at 110°C (h)	EN-14112	Min. 3h	EN ISO 14112	Min. 6h	3.43	3.71	4.24
Viscosity at 40°C (cSt)	D-445	1.9–6.0	EN ISO 3104	3.5–5.0	4.7	4.1	4.9

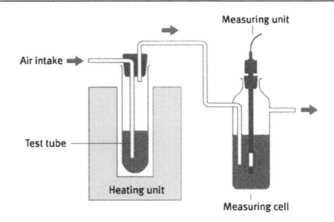

Figure 14.3 Rancimat principle of biodiesel.

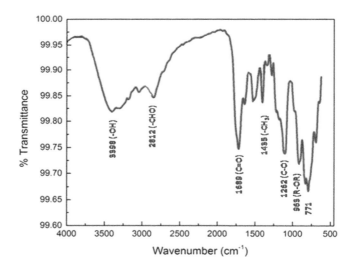

Figure 14.4 FTIR spectrum of *Psidium guajava* extract.

and the percentage inhibition of free radical declines as the amount of PGE is reduced. The observational pattern confirms PGE's antioxidative capability. The reaction pathway of DPPH inhibition with gallic acid, which has been identified as the primary polyphenol in the chemistry of PGE, is depicted in Figure 14.4 (Figure 14.6).

Because the aromatic ring contains a resonance, the phenolic group in its composition yields hydrogen, resulting in the formation of persistent free radicals and hindering the free-radical oxidation chain events in biodiesel.

14.7.4 Overall Phenolic Content (TPE)

A significant component is the phenolic content, which acts as a free radical terminator. The phenolic content of PGE was determined by employing a standardized graph established employing gallic acid. The sample measurement is 8.9 g of GAE/100 g, which is appropriately robust and validates the effect of DPPH and FTIR scavenging PGE results.

14.7.5 Outcome of Guava Extract on the Stability of Oxidation of JBD

The IP of free from PGE biodiesels is 4.24 h for JBD, 3.71 h for PBD and 3.43 h for TGBD, verified by three times employing Rancimat apparatus. Further, more research was important when samples of biodiesel were doped with 100–1000 ppm of PGE as presented in Figures 14.5 and Figure 14.7 (Figure 14.9).

According to EN-14112 requirements, 300 ppm (or more) addition of PGE was sufficient for JBD to meet the least criterion of 6 h and validate the antioxidative capability of *Psidium guajava* extract. However, doping must be equal to or more than 400 ppm for PBD and TG biodiesel to meet the regulatory limit of 6h.

Later adding 1000 ppm of PGE, the JBD, PGE and TGBD were stored for a month. After one month, it was found that the induction period of JBD, PBD and TGBD was decreased to 10.88, 9.77 and 10.06 h as shown in Figure 14.8 (Figure 14.10).

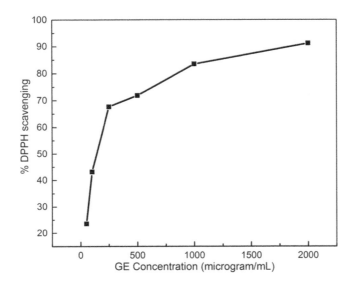

Figure 14.5 DPPH scavenging in the presence of *Psidium guajava* extract.

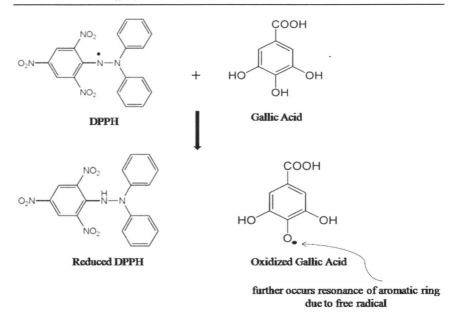

Figure 14.6 Gallic acid's mechanism of DPPH inhibition.

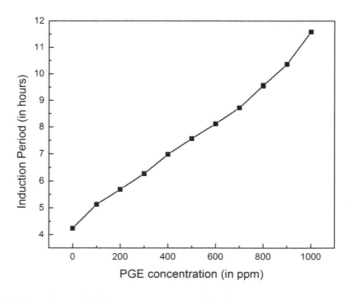

Figure 14.7 The effect of PGE concentration on PBD IP.

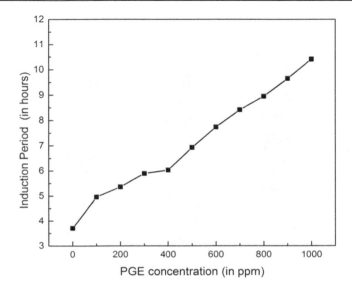

Figure 14.8 The effect of PGE concentration on PBD IP.

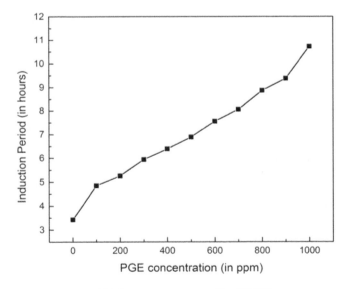

Figure 14.9 Consequence of PGE concentration on IP of TGBD.

14.8 CONCLUSION

Jatropha, *Pongamia*, and *Tectona grandis* biodiesels were tested to enhance the stability of oxidation by adding *Psidium guajava* extract and the results communicated the remarkable potential of PGE as a natural antioxidant.

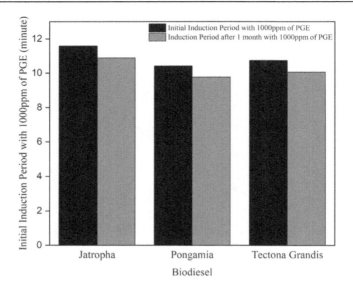

Figure 14.10 Variation of induction period in a month with addition of 1000 ppm of PGE.

Current research discloses that *Psidium guajava* is novel, natural, and reliable which enhances stability of the oxidation of biodiesels. The longer induction duration with increasing additive volume is attributed to the presence of a beneficial functional group in its chemistry. Its antioxidant activities against 2, 2-diphenyl-1-picryl-hydrazyl-hydrate (DPPH) radical were also tested, as was its total phenolic content (11.1 g of Gallic acid equivalent/100 g) using the Folin-Ciocalteau method. The induction duration of clean JBD, PBD and TGBD stayed at approximately 4.24, 3.71, and 3.43 h, respectively, as assessed with 893 professional Biodiesel Rancimat. When doped with 1000 ppm PGE, the induction duration of JBD, PBD and TGBD increases by approximately 11.59 h, 10.41 h, and 10.73 h, respectively. After adding 1000 ppm of PGE, the JBD, PBD, and TGBD were held for one month, and it was discovered that the induction period decreased by roughly 6.16%. Thus, the use of new PGE as an antioxidant for the used biodiesels showed increased resistance to oxidative stability.

REFERENCES

1. Pryde, E. H. (1983). Vegetable oils as diesel fuels: Overview. *Journal of the American Oil Chemists' Society*, 60(8), 1557–1558.
2. Sarin, A. (2012). *Biodiesel Production and Properties*, RSC Publications, Thomas Graham House, Cambridge, UK, ISBN: 978-1-84973-470-7.

3. Fernandez, C. M., Solana, M., Fiori, L., Rodriguez, J. F., Ramos, M. J., Perez, A. (2015). From seeds to biodiesel: Extraction, esterification, transesterification and blending of *Jatropha curcas* oil. *Environmental Engineering and Management Journal*, 14, 2855–2864.

4. Sarin, A., Arora, R., Singh, N. P., Sarin, R., Malhotra, R. K., Kundu, K. (2009). Natural and synthetic antioxidants: Influence on the oxidative stability of biodiesel synthesized from non-edible oil. *Energy*, 34, 2016–2021.

5. Beaver, A., Castano, A. G., Diaz, M. S. (2016). Life cycle analysis of *Jatropha curcas* as a sustainable biodiesel feedstock in Argentina. *Chemical Engineering Transactions*, 50, 433–438.

6. Agarwal, A. K., Khurana, D., Dhar, A. (2015). Improving stability of oxidation of biodiesels derived from Karanja, Neem and Jatropha: Step forward in the direction of commercialization. *Journal of Cleaner Production*, 07, 646–652.

7. Gui, M. M., Lee, K. T., Bhatia, S. (2008). Feasibility of edible oil vs. non-edible oil vs. waste edible oil as biodiesel feedstock. *Energy*, 33, 1646–1653.

8. Sarin, A., Singh, M., Sharma, N., Singh, N. (2017). Prospects of *Tectona grandis* as a feedstock for biodiesel. *Frontiers in Energy Research*, 5. 10.33 89/fenrg.2017.00028

9. Knothe, G. (2007). Some aspects of biodiesel oxidative stability. *Fuel Processing Technology*, 88, 669–677.

10. Cremonez, P. A., Feroldi, M., Jesus Oliveira, C., Teleken, J. G., Meier, T. W., Dieter, J., Borsatto, D. (2016). Oxidative stability of biodiesel blends derived from different fatty materials. *Industrial Crops and Products*, 89, 135–140.

11. Knothe, G., Van Gerpen, J., Krahl, J. (2004). *The biodiesel handbook*. AOCS Press, Champaign, IL, USA.

12. The EN 14214 Standard-Specifications and Test Methods (2008).

13. American Society for Testing and Materials. (2011). *ASTM D6751-11b, Standard Specification for Biodiesel Fuel Blend Stock (B100) for Middle Distillate Fuels*. ASTM International, West Conshohocken, PA.

14. Rizwanul Fattah, I. M., Masjuki, H. H., Kalam, M. A., Hazrat, M. A., Masum, B. M., Imtenan, S. (2014). Effect of antioxidants on stability of oxidation of biodiesel derived from vegetable and animal based feedstocks. *Renewable Sustainable Energy Reviews*, 30, 356–370.

15. Race, S. (2009). Antioxidants- The truth about BHA, BHT, TBHQ and other antioxidants used as food additives. *Tigmor Books*, ISBN: 9781907119002.

16. Lobo, V., Patil, A., Phatak, A., Chandra, N. (2010). Free radicals, antioxidants and functional foods. *Pharmacognosy Reviews*, 4, 118–126.

17. Thaipong, K., Boonprakob, U., Crosby, K., Cisneros- Zevallos, L., Hawkins-Byrne, D. (2006). Comparison of ABTS, DPPH, FRAP, and ORAC assays for estimating antioxidant activity from guava fruit extracts. *Journal of Food Composition and Analysis*, 19, 669–675.

18. Kwee, L. T., Chong, K. K. (1990) *Guava in Malaysia: Production, Pests and Diseases*. Tropical Press Sdn. Bhd, Kuala Lumpur.

19. Mercadante, A. Z., Steck, A., Pfander, H. (1999). Carotenoids from guava (*Psidium guajava* L.): Isolation and structure elucidation. *Journal of Agricultural Food and Chemistry*, 47, 145–151.

20. Musa, K. H., Abdullah, A., Subramaniam, V.. (2015). Flavonoid profile and antioxidant activity of pink guava. *ScienceAsia*, 41, 149–154.

272 Thermal Energy Systems

21. Seo, J., Lee, S., Elam, M., Johnson, S. A., Kang, J., Arjmandi, B. H. (2014). *Study to find the best extraction solvent for use with guava leaves (*Psidium guajava *L.) for high antioxidant efficacy.* Food Science & Nutrition, Published by Wiley Periodicals, Inc., 174–180.

22. KoohiKamali, S., Tan, C. P., Ling, T. C. (2012). Optimization of sunflower oil transesterification process using sodium methoxide. *Scientific World Journal*, 475027. 10.1100/2012/475027

23. Qin, S., Sun, Y., Meng, X., Zhang, S. (2010). Production and analysis of biodiesel from non-edible seed oil of *Pistacia chinensis. Energy Exploration and Exploitation*, 28, 37–46.

24. Surewicz, W. K., Mantsch, H. H., Chapman, D. (1993). Determination of protein secondary structure by Fourier transform infrared spectroscopy: A critical assessment. *Biochemistry*, 32, 389–393.

25. Gorinstein, S., Haruenkit, R., Poovarodom, S., Vearasilp, S., Ruamsuke, P., Namiesnik, J., Leontowicz, M., Leontowicz, H., Suhaj, M., Sheng, G. P. (2010). Some analytical assays for the determination of bioactivity of exotic fruits. *Phytochemical Analysis*, 21, 355–362.

26. Szymanska-Chargot, M., Zdunek, A. (2013). Use of FT-IR spectra and PCA to the bulk characterization of cell wall residues of fruits and vegetables along a fraction process. *Food Biophysics*, 8(1), 29–42.

27. Brand-Williams, W., Cuvelier, M. E., Berset, C. (1995). Use of a free radical method to evaluate antioxidant activity. *Lebensmittel-Wissenschaft und Technologie*, 26, 25–30.

28. Vasco, C., Ruales, J., Kamal-Eldin, A. (2008). Total phenolic compounds and antioxidant capacities of major fruits from Ecuador. *Food Chemistry*, 111, 816–823.

29. Carvalho, A. L., Cardoso, E. A., Rocha, G. O., Teixeira, L. S. G., Pepe, I. M., Grosjean, D. M. (2016). Carboxylic acid emissions from soybean biodiesel oxidation in the EN14112 (Rancimat) stability test. *Fuel*, 173, 29–36.

30. Yuliarita, E., Fathurrahman, N. A., Aisyah, L., Hermawan, N., and Anggarani, R. (2019). Comparison of synthetic and plant extract antioxidant additives on biodiesel stability. In IOP Conference Series: Materials Science and Engineering, 494 (1), 012030. IOP Publishing.

31. Osawa, W. O., Sahoo, P. K., Onyari, J. M., Mulaa, F. J. (2016). Effects of antioxidants on oxidation and storage stability of *Croton megalocarpus* biodiesel. *International Journal of Energy and Environmental Engineering*, 7 (1), 85–91.

32. Yang, J., He, Q. S., Corscadden, K., Caldwell, C. (2017). Improvement on oxidation and storage stability of biodiesel derived from an emerging feedstock camelina. *Fuel Processing Technology*, 157, 90–98.

33. Jain, S., Sharma, M. P. (2011). Long term storage stability of *Jatropha curcas* biodiesel. *Energy*, 36(8), 5409–5415.

34. Chaithongdee, D., Chutmanop, J., Srinophakun, P. (2010). Effect of antioxidants and additives on the stability of oxidation of jatropha biodiesel. *Agriculture and Natural Resources*, 44(2), 243–250.

35. Das, L. M., Bora, D. K., Pradhan, S., Naik, M. K., Naik, S. N. (2009). Long-term storage stability of biodiesel produced from Karanja oil. *Fuel*, 88(11), 2315–2318.

36. Widarti, S., Budiastuti, H., Prasetyani, T., Adhiawardana, S. (2019). Effect of storage towards stability of oxidation and physical properties of biodiesel from palm fatty acid distillate (PFAD). *Journal of Physics: Conference Series*, 1295(1), 012065. IOP Publishing.

37. Romola, C. V., Jemima, M., Meganaharshini, S. P., Rigby, I., Ganesh Moorthy, R., Kumar, S., Karthikumar, S. (2021). A comprehensive review of the selection of natural and synthetic antioxidants to enhance the oxidative stability of biodiesel. *Renewable and Sustainable Energy Reviews*, 145, 111109.

Index

Active beams, 90
Active Oxygen Method (AOM), 256
AHP-TOPSIS, 43
Air-BTMS, 221
Air Conditioning Systems, 93
Air cooling, 222
Air Handling Units (AHU), 96
Air quality, 76
Ambient conditions, 171, 178, 179
Analogue models, 108
Analytic Hierarchy Process (AHP), 46
ANOVA, 207
Antioxidant, 249
Artificial data, 104
Artificial Neural Fuzzy Interface System (ANFIS), 237
Artificial neural network (ANN), 184, 237
ASHRAE, 43, 193

Batteries, 220
Battery pack (BP), 220
Battery Thermal Management System (BTMS), 221
Biodiesel, 231, 249, 251, 257
Biomass, 9, 83
Blend preparation, 234
Body Core, 77
Brake thermal efficiency, 240
Building, 78, 169
Building heat load, 87
Building heating and cooling mode, 183
Building-related illness (BRI), 78
Building Service Systems, 98
Butylated hydroxyanisole (BHA), 260

Cause and effect, 186
Carbon-based composites, 52
Carbon emissions, 17

Central heating systems, 84
Chemical Heat Storage, 18, 23, 24
Classification of Solar Collectors, 32
Classification of TES systems, 19
Climatic conditions, 41
Closed-Air Gap, 79
CO emission, 241
Combustion boilers, 84
Compound Parabolic Collectors, 34
Compression ratio (CR), 232
Computational fluid dynamics (CFD), 233
Confirmation test, 213
Continuous or discrete simulation, 133
Control volume, 107
Convective thermal transfer, 52
Cooling conceptualization, 92
Covid-19, 219, 231
Crystalline configuration, 66

Decentralized Energy Systems, 11
Decentralized Heat Generators, 83
Dehumidification, 172, 178
Dehumidifier, 174, 175
Desiccant cooling, 169
Desiccant wheel, 173, 178
Deterministic or stochastic simulation, 134
Diesel blends, 235
Diesel consumption rate, 233
Diesel engine, 231
Differential Thermal Analysis (DTA), 255
Direct Evaporation, 93
Distribution Systems, 89
District Heating, 87
Domestic hot water (DHW), 91
Dye-Sensitized, 69

Energy analysis, 174
Energy consumption, 2

275

276 Index

Energy conversion, 29
Energy efficiency of buildings, 98
Energy resources, 1
Energy storage systems, 18
Education Institutions, 12
Electric Vehicles (EVs), 219, 220, 221
Electrical efficiency, 166
End heat exchangers, 90
Entropy generation, 175
Evacuated Tube Collectors (ETC), 35
Evaluation of uncertainty, 45
Evaporation rate, 144
Exhaust emissions, 232
Exergy, 177, 178, 179

Fatty Acid Compositions, 257
Fatty Acid Methyl Ester (FAME), 261
Finite Element Method, 127
Fins, 141
Fission, 5
Flat Plate Collector (FPC), 33, 34, 49
Fluorescence Spectroscopy, 254
Fossil fuels, 29
Future Energy Systems, 9

Geographical constraints, 70
Geothermal energy, 6
GHX, 184, 185, 192, 193, 194, 214
Global warming, 17
Gravimetric analysis, 256
Ground-contact building structures, 81
Ground heat exchanger, 98
Green structures, 80
Ground Source Heat Pump (GSHP),
    183, 184, 186, 214

Heat dissipation, 221
Heat exchanger, 207
Heat generation, 220
Heat Generation Technologies, 81
Heat pumps, 86
Heat Recovery Units (HRU), 96
Heat storage materials (HSMs), 141
Heat transfer liquid, 33
Heat Transfer Mechanisms, 78
Heliostat Field Collector, 37
HEV, 219, 220, 221, 224, 227, 230
Homogeneous structures, 79
Hot thermal fluids (HTF), 19
HVAC, 5
Hybrid-built cooling system, 170
Hybrid configuration, 68
Hydrogen, 231

Hydrogen fuel blending concentration
    (HBC), 233
Hydrogenating, 10
Hydronic Systems, 89

Ignition delay (ID), 232
Industry 4.0, 61
Internal combustion engine (ICE),
    220, 221

James Watt, 2
Jet Fuel Thermal Oxidation Tester
    (JFTOT), 257

K-type thermocouples, 144

Latent Heat Storage (LHS), 23
Linear Fresnel Reflector, 36
Liquid-BTMS, 223
Lithium-ion battery (LiB), 219, 220, 226

Main effect analysis, 207
Mass flow, 175
Mapping technique, 128–9
Mathematical models, 109
Mechanical Cooling Systems, 92
Mechanical ventilation, 95
Methodology, 45, 134, 157, 160, 186
Modelling, 103
Moisture Removal Rate (MRR), 174
Multi-criteria decision-making
    (MCDM), 43
Municipalities, 13

Nanofluids, 41, 43, 44, 160, 157, 160
Nanoparticles, 159, 231, 234
Natural antioxidant, 259
Natural ventilation, 93
Net-Zero Energy Buildings (nZEB), 73,
    75, 76
NOx Emission, 243
Numerical Models, 115

Optimal Levels, 213
Optimization, 103, 136, 187
Optimization of Thermal Systems, 136
Optimum Operating Parameters, 183, 214
Orthogonal Array, 189
Overall Phenolic Content (TPE), 266
Oxidation, 249, 251, 253

Pandemic, 219
Parabolic Dish Reflector, 36

Index   277

Parabolic Through Collector, 35
Parameters calculation, 192
Passive air cooling, 222
PCM, 23, 27, 141, 142, 155, 156, 158, 159, 220
PCM-BTMS, 222
Performance evaluation, 231
Performance Parameters for TES, 25
Physical Models, 131
Piezoelectric, 235
Polymer and Organic Configuration, 68
Pollution, 169, 219
Productivity, 150
Psidium guajava, 249, 261, 263
PSS absorber basin, 144
PV, 161
PV/T, 156, 157, 160, 167
Pyramid solar stills (PSS), 143

Radiators, 90
Reactivation air, 175
Recirculation, 170
Regeneration, 171
Regulate Heat Transmission, 78
Renewable Energy, 157, 169
Renewable Energy Technologies, 12
Response Surface Methodology (RSM), 236
Reynolds number, 222

Scavenging, 262
Second Law, 169
Semiconductor, 69
Sensible Energy Storage Technologies, 20
Sensible heat storage, 18, 21
SHS material, 23
Sick building syndrome (SBS), 77
Silicon-based technology, 65
Simulation, 103
Smart applications, 7
Smart thermal grids, 9
S/N ratio, 198
Socio-cultural factors, 10
Solar cell materials, 65
Solar collectors (SC), 42
Solar Energy, 5, 37, 38
Solar energy conversion, 30
Solar energy system, 8, 27
Solar power system, 27
Solar PV, 59, 63, 64, 69
Solar PV Concentrated Technology, 67

Solar PV/T System, 155
Solar simulator, 158
Solar still, 141
Solar System TES, 25, 26
Solar thermal collectors, 86
Solar Thermal System, 62
Solid storage materials, 22
Sources of thermal energy, 3
Space cooling and heating, 199, 208
Space heating systems, 63
Space setup, 92
Stability of oxidation, 263
Steady or dynamic simulation, 133
Sun tracking collectors, 35
Synthetic antioxidant, 260
System configuration, 157
System operation, 171

Taguchi Method, 187
TES, 19, 24, 103
TES process layout, 18
TES Systems, 20, 21
Tesla, 223, 224
THD (Total Harmonic Distortion), 70
Thermal and economic analysis, 145
Thermal and electrical management, 155
Thermal borehole resistance, 206, 210
Thermal comfort, 76
Thermal conductivity, 166, 198
Thermal cooling systems, 170, 171, 179
Thermal efficiency, 85
Thermal efficiency enhancement, 141, 150
Thermal energy applications, 73
Thermal energy resources, 7
Thermal energy storage (TES), 18, 26
Thermal energy storage (TES) technologies, 20
Thermal Management Systems, 219
Thermal receiver, 37
Thermal recovery, 98
Thermal systems, 169
Thermal technologies, 64
Thermally Activated Building Structures, 91
Thermochemical heat storage system, 24
Thermochemical storage (TCS), 24
Thermodynamic, 177
Thermogravimetric Analysis (TGA), 255

Thermostat, 172
Thin Film Technology, 66
Three-dimensional problem, 107
TOPSIS Assignment, 47
Transient, 112
Transportation, 220

UBHC Emissions, 242
Unburnt hydrocarbons (UBHC), 242
Utilization of solar energy, 31

Variation of Operating Conditions, 45
Varying Injection Pressure (VIP), 231
Ventilated Air Layer, 79
Ventilation Energy Efficiency, 98
Ventilation Systems and Management, 93
VCR, 172

Waste Heat Energy, 6
World Consumption of Primary
    Energy, 4